Frontiers of Plasmon Enhanced Spectroscopy Volume 1

ACS SYMPOSIUM SERIES **1245**

Frontiers of Plasmon Enhanced Spectroscopy Volume 1

Yukihiro Ozaki, Editor
Kwansei Gakuin University
Sanda, Hyogo, Japan

George C. Schatz, Editor
Northwestern University
Evanston, Illinois

Duncan Graham, Editor
University of Strathclyde
Glasgow, United Kingdom

Tamitake Itoh, Editor
National Institute of Advanced Industrial Science and Technology
Kagawa, Japan

American Chemical Society, Washington, DC

Distributed in print by Oxford University Press

Library of Congress Cataloging-in-Publication Data

Names: Ozaki, Y. (Yukihiro), editor.
Title: Frontiers of plasmon enhanced spectroscopy / Yukihiro Ozaki [and three others], editor.
Description: Washington, DC : American Chemical Society, [2016]- | Series: ACS symposium series ; 1245, 1246 | Includes bibliographical references and index.
Identifiers: LCCN 2016054894 (print) | LCCN 2016055672 (ebook) | ISBN 9780841232013 (v. 1) | ISBN 9780841232037 (v. 2) | ISBN 9780841232006 (v. 1) (ebook) | ISBN 9780841232020 (v. 2) (ebook)
Subjects: LCSH: Raman spectroscopy. | Spectroscopic imaging. | Plasmons (Physics)
Classification: LCC QC454.R36 F75 2016 (print) | LCC QC454.R36 (ebook) | DDC 543/.57--dc23
LC record available at https://lccn.loc.gov/2016054894

Distributed in print by Oxford University Press

Foreword

The ACS Symposium Series was first published in 1974 to provide a mechanism for publishing symposia quickly in book form. The purpose of the series is to publish timely, comprehensive books developed from the ACS sponsored symposia based on current scientific research. Occasionally, books are developed from symposia sponsored by other organizations when the topic is of keen interest to the chemistry audience.

Before agreeing to publish a book, the proposed table of contents is reviewed for appropriate and comprehensive coverage and for interest to the audience. Some papers may be excluded to better focus the book; others may be added to provide comprehensiveness. When appropriate, overview or introductory chapters are added. Drafts of chapters are peer-reviewed prior to final acceptance or rejection, and manuscripts are prepared in camera-ready format.

As a rule, only original research papers and original review papers are included in the volumes. Verbatim reproductions of previous published papers are not accepted.

ACS Books Department

Contents

Indexes

Preface

More than four decades have passed since surface-enhanced Raman scattering (SERS) was discovered. In today's world SERS has been established as a plasmon-based spectroscopy with ultra-high sensitivity and versatility at the forefront of the developments in plasmonics. SERS has been developing with the advances in nanoscience and nanotechnology. The "SERS world" has grown up markedly for the last 20 years or so, and recently the wider concept of, plasmon-enhanced spectroscopy was born. Plasmon-enhanced spectroscopy contains not only SERS but also tip-enhanced Raman scattering (TERS), surface-enhanced infrared absorption (SEIRA), surface-enhanced fluorescence (SEF), and more. Through these novel spectroscopies various amazing properties of plasmons have become known, providing new exciting research fields.

In Pacifichem 2015, held in December 2015 in Hawaii, we had a symposium titled "Frontiers of Plasmon Enhanced Spectroscopy." This book is concerned with the symposium, although it is not its proceedings. Thus, the collection is based on the above symposium, and most of the contributors to this book were its invited speakers. One of the main purposes of the book is to convey the enthusiastic discussion on plasmon-enhanced spectroscopy at the symposium to the scientific community.

The book reports leading-edge advances in the theory of plasmonic enhancement and application of plasmon-enhanced spectroscopy to biology, chemistry, physics, materials science, and medicine. Many books have been published about SERS, but this may be the first time that a book on a wide area of plasmon-enhanced spectroscopy has ever been published. The book consists of two volumes; the first volume contains the introductory review by George Schatz followed by eight chapters, which are mainly concerned with SERS. The second volume discusses TERS, SEIRA, and other topics related to plasmon-enhanced spectroscopy.

Last but not at least, we hope the readers not only learn a great deal about the-state-of-the-art of plasmon-enhanced spectroscopy but also enjoy this book. We will be most grateful if the book becomes a trigger to open further new exciting fields in surface-enhanced spectroscopy.

Yukihiro Ozaki
Kwansei Gakuin University
Sanda, Hyogo, Japan

George C. Schatz
Northwestern University
Evanston, Illinois

Duncan Graham
University of Strathclyde
Glasgow, United Kingdom

Tamitake Itoh
National Institute of Advanced Industrial Science and Technology
Kagawa, Japan

Chapter 1

Review of Plasmon-Induced Hot-Electron Dynamics and Related SERS Chemical Effects

Rebecca L. Gieseking, Mark A. Ratner, and George C. Schatz*

Department of Chemistry, Northwestern University, 2145 Sheridan Road, Evanston, Illinois 60208, United States
***E-mail: g-schatz@northwestern.edu**

The plasmonic enhancement of photochemistry and spectroscopy is of broad interest in chemistry, physics and engineering. Here, we review the fundamental processes and time scales associated with plasmonic excitation and decay processes, including studies of the interactions of plasmons with nearby molecules or semiconductor particles as involves energy and electron transfer between the two moieties. This work includes interactions that contribute to photovoltaic or photocatalytic processes and to the chemical enhancement mechanism in surface-enhanced Raman scattering. Understanding these processes is critical to the design of more efficient plasmon-enhanced devices and spectroscopic applications.

Introduction

Plasmon-enhanced chemistry has recently been of significant interest in the research community as many of the fundamental processes involved are poorly understood. Since plasmonic metal nanoparticles have very large absorption cross-sections, they are potentially useful as antennas for capturing and converting solar energy. As a result, there have been several reports of plasmon-molecule or plasmon-semiconductor interactions for converting solar energy into electrical or chemical energy via photovoltaic or photocatalytic processes, often involving the production and decay of hot-electrons (*1–4*). In addition, plasmon excitation has been widely used for spectroscopic applications, in particular in surface-enhanced Raman spectroscopy (SERS) (*5–7*). The underlying physics of the plasmon enhancement mechanisms leading to SERS and hot electron formation are intimately related, and indeed often the same substrates are used in both studies.

To better develop these types of applications, it is critical to understand the fundamental processes involved in the excitation and decay of plasmons and in the interactions of the plasmons with molecules or semiconductors, as summarized in Figure 1. Following the plasmon excitation, the coherent excitation dephases into individual charge carriers, typically within tens of fs. The initial non-thermal distribution of energetic charge carriers relaxes to a hot Fermi-Dirac thermal distribution on the order of hundreds of fs, and the excess electronic energy is transferred to vibrational energy on the order of a few ps. On a longer time scale, the vibrational energy within the plasmonic structure dissipates to the environment. Interactions of the plasmonic nanostructures with nearby organic molecules or semiconductors can introduce new decay pathways involving the transfer of energy or charge carriers between the two moieties. We note that while the excited electrons and holes before thermalization are commonly referred to as hot carriers, this term may be somewhat misleading as an effective temperature cannot be defined for the non-thermal energy distribution and a term such as 'energetic carriers' may be more precise. However, we primarily use the more common term 'hot carriers' in this paper.

In this review, we will describe the fundamental processes involved in each of these steps, focusing on how changes to the nanoparticle structure and environment affect the rates and energy distributions involved. The primary emphasis is on theoretical studies involving classical and quantum mechanical descriptions of the plasmon but we also compare to a variety of experimental results. Connections with SERS theory will also be discussed, especially concerning the chemical enhancement effect.

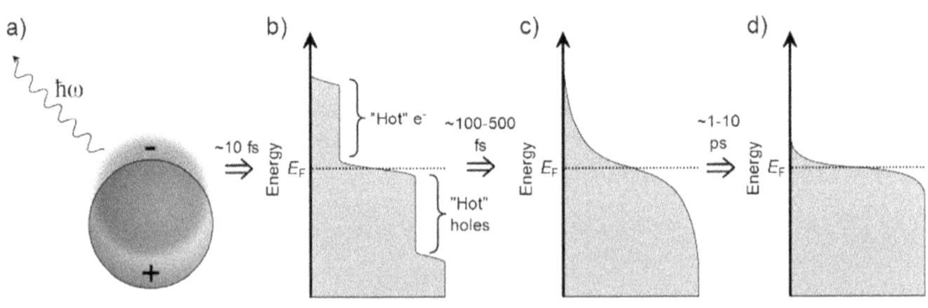

Figure 1. Schematic of the plasmonic decay process. (a) Radiative decay of the plasmon. Non-radiative decay processes including (b) dephasing into energetic electrons and holes, (c) thermalization of hot carriers, and (d) relaxation of hot carriers into phonons.

Basic Description of Plasmons

Applications in plasmonics typically involve metal surfaces or films which support surface plasmon resonance (SPR) states or nanostructures of various shapes which have localized surface plasmon resonances (LSPR). In a classical electrodynamics picture, the plasmon is a coherent oscillation of all conduction-band electrons in the material as shown in Figure 2a. Mie theory is a solution to Maxwell's equations for a sphere that describes the total extinction, including absorption and scattering, and therefore the consequences of plasmon excitation on optical properties (*8*); however, simple analytical solutions to Maxwell's equations have only been found for spherical and spheroid particles (*8*, *9*). For more complex shapes, numerical methods such as the finite-difference time-domain (FDTD) method (*10*) are typically used, where the material is represented on a grid and the electric and magnetic fields are propagated in time according to Maxwell's equations. The oscillation of the conduction electrons in response to the electric field of light creates strong localized enhancement of the electric field near the plasmonic metal surface (Figure 2b) (*11*). The electric-field enhancement is particularly pronounced near 'hot spots' on the nanostructure, which may include sharp points or defects on the surface of individual nanoparticles (*12*) or few-nm gaps between two nanoparticles (*13*, *14*). Although plasmonic resonances may involve dipolar, quadrupolar, or higher multipolar charge oscillations (*11*), our primary interest is in the dipolar resonance, which typically has the strongest absorption.

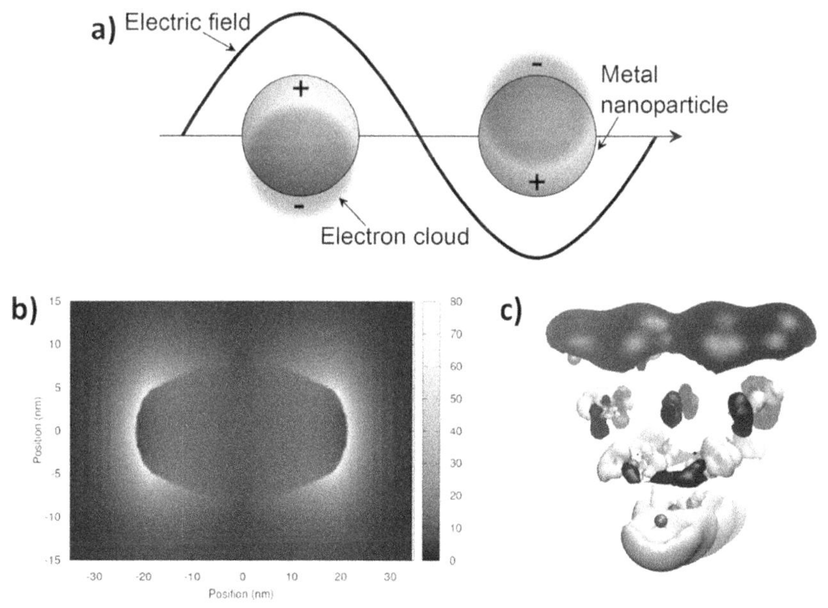

Figure 2. (a) Diagram showing plasmon oscillations for a metal nanosphere including displacement of the conduction electrons that leads to local field enhancement and (b) electric field contours for an Au bipyramid in response to the electric field of light (figure courtesy of Craig Chapman) (15), and (c) INDO/SCI transition density of the plasmonic excited state in a tetrahedral Ag_{20} *cluster (16).*

More recently, a quantum mechanical description of the plasmonic excitations in small nanoclusters has been developed using TD-DFT (*17–19*) and CI (*16*, *20*) approaches. In these models, the collectivity of the plasmonic state is described in terms of a coherent linear combination of several to many single particle-hole excitations that contribute additively to the transition dipole moment, resulting in a large dipolar transition density as shown in Figure 2c (*16*, *17*). The dominant transitions in the plasmonic state are sp $\rightarrow$ sp intraband transitions. Analogously, real-time simulations reveal relatively narrow absorption peaks corresponding to large dipole moment oscillations (*19*, *21*, *22*). The excitations involved in the plasmonic state typically span a broad distribution of electron and hole energies (*19*, *21*). Of the possible linear combinations of these excitations, the plasmonic state tends to be one of the higher energy eigenstates (*20*); equivalently, most of the single particle-hole excitations comprising the plasmon are lower in energy than the plasmonic state (*21*). The size dependence of the quantum mechanical plasmonic energy for small clusters is consistent (when extrapolated to >10 nm particles) with the classical electrodynamic results (*18*).

Plasmonic Decay Processes

Absorption, Scattering, and Radiative Decay Processes

In small spherical Au nanoparticles, absorption scales with volume whereas the scattering scales with the square of volume, as predicted by Mie theory (*23*). As a result, absorption is dominant for small particles, while elastic scattering is the dominant interaction for large particles. The size at which scattering becomes the dominant process is between 50 and 100 nm for Au nanospheres depending on the wavelength considered (*23*) and similarly ranges from 50 nm in Ag nanospheres (*24*) to 110 nm in Ag nanodisks (*25*).

In addition to absorption and scattering, excited states can decay by one-photon luminescence. This is a particularly inefficient process for flat metal surfaces with each absorbed photon having a probability of 10^{-10} of being re-emitted (*26*). Although this process may be enhanced in nanostructures, it is still a highly inefficient process, with efficiencies ranging from 10^{-6} for nanospheres (*27*) to 10^{-4} for nanorods (*28*).

Dephasing

Since radiative plasmon decay is typically inefficient, the primary decay route is a multi-step non-radiative process. Although plasmonic excitation is initially coherent, the first step of the non-radiative decay is dephasing into individual electrons and holes on a time scale ranging from less than 10 fs to a few tens of fs. Since these charge carriers are at energies significantly higher than would be expected in a thermal distribution at the temperature at which the experiments are performed, they are commonly referred to as 'hot' electrons and holes. In an electrodynamics picture, this decay occurs primarily via Landau damping (*29*), where the coherent electromagnetic wave loses energy by scattering electrons moving at velocities similar to the plasmon phase velocity into states above the

Fermi level. In this process, the conservation of momentum only allows a small transfer of energy during each scattering event (*30, 31*). Similar scattering can occur at interfaces or defects or via interactions with phonons; these processes also allow for scattering into much higher energy levels (*31*). In some cases, radiative processes in larger nanostructures can also accelerate the dephasing (*32*). Quantum mechanically, the coherent plasmonic state can interact with single particle-hole excitations resonant with the plasmon excitation energy, leading to transfer of the energy to single excitations (*21*).

Since electron-phonon scattering can accelerate dephasing at room temperature, temperature-dependent measurements of the linewidth of the plasmonic absorption can be used to extract the intrinsic material dephasing rate. These measurements suggest that electron-phonon interactions contribute roughly 30% to the dephasing rate at room temperature and give intrinsic damping rates of 15 fs for Au and 25 fs for Ag (*33*). This estimate of the intrinsic damping is consistent with a Drude model fit to the bulk dielectric constant of Au, which estimated the damping time to be 14 ± 3 fs (*34*). The contribution of electron-phonon coupling is consistent with estimates using ab initio molecular dynamics simulations, which suggested the electron-phonon mechanism alone would cause dephasing in 30-40 fs and thus likely contributes 25-30% to the linewidth (*35*). The dephasing time is also strongly dependent on the wavelength at which the excitation occurs. For an 14-18 nm Au nanoparticle, the dephasing time decreases from ~17 fs to ~5 fs as the photon energy increases from 1.45 eV to 2.15 eV, primarily due to accessibility of interband transitions from the d band into the sp conduction band starting around 1.8 eV (*36*).

Recent single-particle femtosecond-resolved measurements have examined the dependence of dephasing times on the structural properties of individual nanoparticles. For Au nanorods and nanocrosses on the scale of tens of nm, the dephasing time depends much more strongly on the photon energy than on the exact particle size or shape (*37*); a similar weak dependence on the specific particle structure was seen for 14-18 nm Au nanoparticles (*36*). In contrast, individual Ag nanoparticles and nanoclusters on the order of 100 nm show several possible dephasing patterns that vary by particle (*38*). Although spherical nanoparticles typically have a simple exponential decay corresponding to a single dephasing time, slight ellipticity of the particles leads to two distinct rates of dephasing with some beating between the two frequencies.

Interactions of the nanoparticle with its environment allow additional contributions to the dephasing, known as chemical interface damping. These interactions accelerate dephasing by allowing transient charge transfer into surface states, thereby increasing the total electron scattering rate (*39, 40*). These effects are particularly significant in smaller nanoparticles, where surface effects can play a larger role. For 6-13 nm Au nanoparticles on sapphire, the dephasing can be accelerated from a bulk-like time of 15 fs up to 9 fs, depending on the particle size and shape (*41*). Similarly, interactions with small Ag nanoparticles with sapphire substrates (*42*) or with SO_2 (*43*) shorten the dephasing time, with dephasing times as short as 2.6 fs measured for 2 nm particles on sapphire (*42*). Organic ligands used to stabilize Ag NPs can have similar effects on the dephasing time, particularly if they introduce low energy surface states (*44*).

Initial "Hot" Electron/Hole Distribution

Because the electrons and holes produced immediately after dephasing are highly energetic, the initial distribution of electrons and holes is non-thermal and far from the Fermi-Dirac distribution expected for charge carriers at a specific temperature. The exact initial energetic distribution depends on the geometry and material properties of the system. As mentioned previously, the conservation of momentum in the Landau damping process limits transitions to states that are close in energy; this results in a large population of hot electrons and holes within a few tenths of an eV of the Fermi level (*30*, *31*). We note that thermal energy at room temperature is approximately 0.025 eV, so even these relatively low-energy carriers are significantly more energetic than the unexcited thermal distribution of carriers. Interactions with interfaces or with defects or phonons are required to relax these selection rules and produce hot electrons with higher energies. When the selection rules are relaxed, the carriers may have energies up to the plasmon frequency (*30*). Similarly, hot spots where the local electric field and the plasmonic absorption are enhanced can produce more energetic charge carriers (*45*).

Many experimental studies of the decay dynamics of plasmons have observed that approximating the initial hot electron distribution as a thermal distribution could not fully explain the experimental results. In these cases, approximating the initial hot electron distribution as a linear combination of a flat distribution from the Fermi level to the excitation energy and a hot thermal distribution improves the fit of data with experiments (*46–48*). Similarly, time-resolved two-photon photoemission experiments on thin films or bulk plasmonic metals also typically show a thermal-like peak at low energies followed by a relatively flat distribution up to the excitation energy (Figure 3a) (*49–51*).

Simulations using a Fermi gas approach suggest that plasmonic nanoslabs produce hot-electron distributions with a low-energy spike followed by a relatively flat distribution of electrons up to the excitation energy, consistent with experimental observations. For large slabs, the low-energy carriers dominate; for slabs on the order of 10 nm, the distribution of carriers is much flatter, as shown in Figure 3b (*52*, *55*). Similar hot electron distributions are seen for few-nm spheres and cubes (*52*), and a kinetic DFT formalism gives consistent results (*56*). A free-electron model of hot electron generation in Ag nanospheres suggests that the appearance of discrete energy levels for small nanostructures and longer carrier lifetimes contribute to the production of high-energy carriers (Figure 3c) (*53*). In small nanoclusters, the finite number of discrete energy levels significantly limits the electron and hole energies. For the Ag_{55} cluster, only a few specific electron-hole pairs at the plasmon energy are excited as hot carriers (*21*).

The distribution of hot carriers can also depend strongly on the energetic accessibility of d $\rightarrow$ sp interband transitions; the onset of interband transitions occurs at lower energies and plays a more significant role in the plasmonic processes for Au than for Ag. When the excitation energy is low enough that interband transitions are inaccessible, the energy distributions of electrons and holes are relatively flat, as seen in free-electron models and the excitation energy nearly equally distributed between hot electrons and holes. However, when interband transitions are energetically accessible, most of the hot carriers are produced from these interband transitions, and the hot holes are considerably more energetic than the hot electrons as depicted in Figure 3d (*54*, *57*).

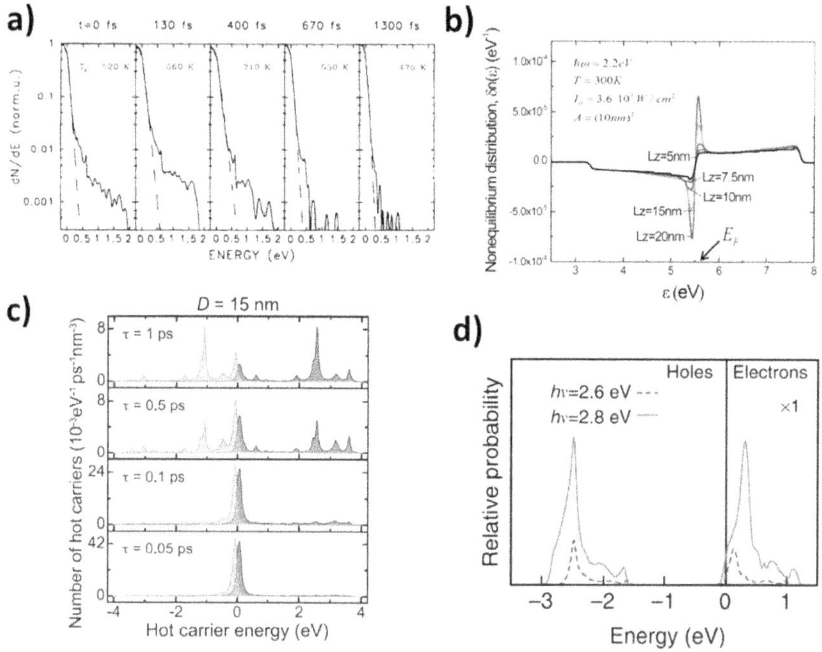

Figure 3. Energetic distributions of hot carriers. (a) Hot electron distribution in bulk Au following excitation at 1.84 eV as measured via time-resolved photoemission. Reproduced from (49). Copyright 1992, American Physical Society. (b) Theoretical hot carrier distribution for metal nanoslabs with thicknesses of 5-20 nm using a Fermi gas approximation. Reproduced from (52). Copyright 2014, American Chemical Society. (c) Distribution of hot electrons and holes in a 15 nm nanosphere as a function of hot carrier lifetime τ. Reproduced from (53). Copyright 2014, American Chemical Society. (d) Hot carrier distribution in Au showing the dominance of d → sp interband transitions. Reproduced from (54). Copyright 2014, Nature Publishing Group.

Thermalization of Hot Carriers

Although the initial distribution of hot electrons and holes is highly non-thermal, electron-electron scattering interactions redistribute the excess energy such that the carriers reach a Fermi-Dirac distribution of electron and hole energies corresponding to a high temperature. In a free electron gas model, a hot electron loses on average half of its energy per scattering event (*58*), so several scattering events are needed to cool an electron from an optical energy of a few eV to a thermal energy. Although the lifetimes of hot carriers at each energy are on the order of tens of fs, the total time for the non-thermal electron distribution to decay to a thermal distribution is typically several hundred fs (*49*, *50*, *59*).

The hot-electron lifetimes can be measured using time-resolved two-photon photoemission, since the hot electrons generated by the pump will be photoemitted if the probe excites them to an energy above the Fermi level (*60*). Size effects are important to consider when interpreting the experimental results, since charge carriers may be transported away from the pulse region in bulk metals or thin films, leading to the appearance of faster decay despite similar scattering rates (*51*, *61*). Femtosecond-resolved reflectivity measurements suggest that the total relaxation time to a thermal distribution is on the order of 350 fs for Ag (*59*) and 500 fs for Au (*62*). The faster relaxation in Ag is likely due to a smaller contribution of the d-band electrons in screening electron-electron interactions. The decay of highly energetic electrons on the order of 1 eV in Au nanostructures is faster than can be resolved using a 130-fs laser pulse (*45*).

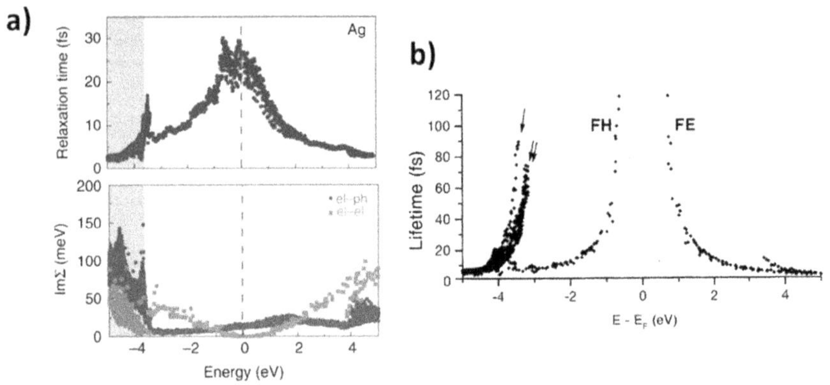

Figure 4. (a) Lifetimes (top) and electron-electron and electron-phonon scattering rates (bottom) for bulk Ag. Reproduced from (57), Copyright 2015, Nature Publishing Group. (b) Computed electron lifetimes including only electron-electron scattering for bulk Ag. Reproduced from (63), Copyright 2001, American Physical Society.

The electron-electron scattering rate and thus the hot carrier lifetimes are strongly dependent on carrier energy, as shown in Figure 4. Within a GW formalism, the hot carrier lifetime is inversely proportional to the imaginary part of the self-energy. Charge carriers close to the Fermi level have low electron-electron scattering rates and thus long lifetimes (*57*); these scattering rates imply that carriers within 1 eV of the Fermi level would have lifetimes of over 100 fs in the absence of other decay mechanisms (*63*). The electron-electron scattering rate increases for carriers further from the Fermi level, which reduces the lifetime to less than 10 fs for hot electrons above roughly 3 eV (*63*). Holes near the top of the d-band have significantly reduced electron-electron scattering than sp-band holes of comparable energy, increasing their lifetime (*57*, *63*). The experimental thermalization dynamics show similar effects of carrier energy and interband transitions, where the hot carrier lifetimes are generally shorter for higher-energy carriers but slightly increase near the onset of interband transitions in Au (*51*). The initial thermalization dynamics in Au are primarily related to the interband transitions, which can decay via Auger recombination on the scale of

70 fs, where the recombination of a d-hole with an electron transfers energy to a conduction band electron (*64*).

In Au nanoparticles larger than about 10 nm, size effects have no noticeable effect on the total thermalization time (*48*). However, for smaller nanoparticles, surface effects and chemical interface effects can have a significant effect. For Ag nanoparticles in a $BaO\text{-}P_2O_5$ or Al_2O_3 matrix, the thermalization is significantly accelerated for particles smaller than 5 nm, with thermalization in 2 nm particles occurring a factor of two faster than in the bulk (*65*). In contrast, the thermalization in 4 nm Au nanoparticles in TiO_2 is slowed to as long as 800 fs, significantly longer than the bulk thermalization time (*66*).

Relaxation of Hot Carriers

A second mechanism of hot carrier relaxation is transfer of the excess electronic energy to vibrational motion of the nuclei via electron-phonon interactions. Since the vibrational modes, particularly for heavy atoms, are at low frequencies, each scattering event can only transfer a relatively small amount of energy to the vibrational modes.

The simplest model for electron-phonon scattering is the two-temperature model (*67*), where the two temperatures refer to a 'hot' thermal distribution for the electronic system and a cooler thermal distribution for the nuclear motion. In this model, it is assumed that the time scale relaxation of the electrons to a thermal distribution is fast enough that it can be separated from electron-phonon coupling on a longer time scale. Although this model is still widely used (*49*, *62*, *68–71*), its applicability depends on the detailed dynamics of the system of interest, since typical time scales are on the order of a few hundred fs for thermalization and a few ps for electron-phonon cooling. A variety of experimental studies of the rates of electron-phonon scattering have shown that a non-thermal initial electron distribution is required to adequately model the overall decay process (*46–48*).

Ab initio calculations of bulk Au and Ag show that the electron-phonon scattering rate is comparable to that of electron-electron scattering, and electron-phonon scattering is the dominant decay process for carriers close to the Fermi level and for d-holes (Figure 4a) (*57*). Unlike electron-electron scattering, the electron-phonon scattering rate is relatively constant for all carriers within the sp band but is greatly accelerated for d-band carriers. For small metal clusters, AIMD simulations suggest that the majority of the energy transfer involves low-energy vibrational modes with frequencies less than 100 cm^{-1} (*72*). The cooling of the electrons is completed within approximately 3 ps for Ag_{68} (*72*) and 6.8 ps for Au_{55} (*73*).

In Au and Ag films, the time scale of relaxation can be measured using time-resolved reflectivity. The hot carriers relax and equalize their temperature with the lattice vibrations within less than 2 ps, only a few times longer than the electron thermalization time (*47*, *74*). Transient absorption measurements of Au nanoparticles larger than 2-3 nm yield relaxation times comparable to the bulk rate (*48*, *68*, *75*). However, high laser intensities create a higher-temperature electronic distribution, which can double the cooling time (*75*).

Interactions of the nanoparticles with their environment can also change the electron-phonon coupling dynamics. In arrays of Au nanoparticles, interactions of the nanoparticles with water rather than air accelerate the electron cooling from 4.1 ps to 2.6 ps (*76*). For Au nanoshells, organic ligands can accelerate the relaxation from 2.8 ps for the bare particles in water to 1.7-2.8 ps depending on the identity of the ligand; the relaxation rate tends to be faster for molecules that strongly chemisorb and induce large surface dipole moments (*77*). Thiol ligands on 3.5 nm Au nanoparticles similarly increase the electron-phonon coupling rate but also increase the electronic heat capacity relative to amine ligands; these two effects partially cancel but increase the hot electron lifetime by 20% (*69*).

Heat Dissipation

After the electronic energy is transferred to vibrational modes, the heat is transferred to the environment and dissipates. The time scale for heat dissipation is very strongly size dependent, from 10 ps for 4 nm Au nanoparticles to 400 ps for 50 nm nanoparticles (*78*). Since the dissipation for small nanoparticles is only slightly longer than the thermalization time, the time scales may not be fully separable (*78*). This dissipation can cause significant local heating, which has been used for a variety of biological and chemical applications (*79*).

Chemical Processes Involving Plasmons

To this point, we have focused primarily on the excitation and decay processes involving the plasmonic nanostructure itself and only discussed the environment to the extent that those interactions affect processes that take place primarily in the nanoparticle. To harness the plasmonic structures for photochemical or photosensing applications, it is critical to understand how the nanostructures interact with nearby molecules or semiconductors. Since the plasmonic decay is a multi-step process that takes place over several time scales, a variety of plasmon-environment interactions can occur at different stages throughout that process.

As described earlier, the coherent oscillation of the conduction electrons in the plasmonic nanostructure creates enhancements of the local electric field of light by factors of many orders of magnitude close to the nanoparticle surface (*11*). This near-field enhancement has been widely used to enhance a variety of molecular properties. Molecular absorption can be strongly enhanced, leading to higher efficiencies of photovoltaic (*80*, *81*) or photocatalytic (*4*, *82*, *83*) processes. Molecular fluorescence can be enhanced by factors of up to 100 at distances greater than 10 nm but is typically quenched at distances less than 5 nm (*84*, *85*), In addition, the electromagnetic mechanism is widely considered to be the primary enhancement mechanism in surface-enhanced Raman spectroscopy (SERS), enabling detection of single molecules in hot spots where the local electric fields are most strongly enhanced (*6*, *7*, *86*, *87*). Because these processes where the involvement of the plasmon is limited to 'passive' field enhancement have been reviewed extensively (*4*, *6*, *84*, *88*), we focus here on processes that involve excitation of the plasmonic states, resulting in transfer of energy or charge between the plasmonic nanostructure and the surrounding molecules or semiconductor.

Plasmon-Exciton Interactions

At very short time scales, the plasmon may interact with excitonic excited states on nearby molecules or semiconductor particles. When the plasmonic and excitonic states are coupled and nearly in resonance, the system may form hybrid plasmon-exciton excited states (Figure 5a). Since the plasmonic absorption is typically significantly broader than the excitonic absorption, this typically appears in the absorption spectrum as a Fano lineshape (*89–91*). The shape and intensity of the Fano lineshape are affected by the exact energy difference and coupling between the two moieties. Strongly coupled plasmon-exciton states have been modeled using a variety of computational approaches, including fully quantum mechanical approaches (*92*, *93*), mixed quantum-classical approaches (*94*), and classical approaches (*95*, *96*). The coupling between the plasmon and the molecule can lead to energy transfer between the two moieties with time constants as fast as 10 fs (*91*, *96*) or to electron transfer from the molecule into the metal (*95*).

Coupling between the plasmonic nanoparticle and a molecule can also lead to resonance energy transfer to the molecule, in a process analogous to Forster resonance energy transfer (FRET) as depicted in Figure 5b. This process has been referred to in various studies as either plasmon RET (PRET) (*97–100*) or plasmon-induced RET (PIRET) (*101*, *102*). Unlike in traditional FRET, RET involving plasmons must occur on a very short time scale before dephasing is complete, since the oscillator strength is large only for the coherent plasmonic state but not for the individual electron-hole pairs (*101*). The short time scale also means that there is no Stokes shift from the absorption peak, so energy can be transferred from the plasmon to molecules with absorption slightly blue-shifted from the center of the plasmonic absorption peak (*101*, *103*).

In systems of nanoparticles coupled with metal ions or dimers, RET can be confirmed by exciting the plasmon and detecting photoluminescence corresponding to the ion or dimer (*98*, *100*, *104*). Similar fluorescence enhancement at plasmonic absorption wavelengths is seen between Au and CdSe nanoparticles (*105*). In nanoparticles interacting with biomolecules, the RET can be detected by a decrease in scattering at the wavelength at which the molecule absorbs (*97*, *99*). For nanoparticles interacting with semiconductors, energy transfer can be distinguished from electron transfer by comparing the effect of different semiconductors; TiO_2 can only accept electrons from Au nanoparticles because the band gap is significantly larger than the plasmonic absorption energy, whereas Cu_2O only accepts energy because there is no driving force for electron transfer (*106*).

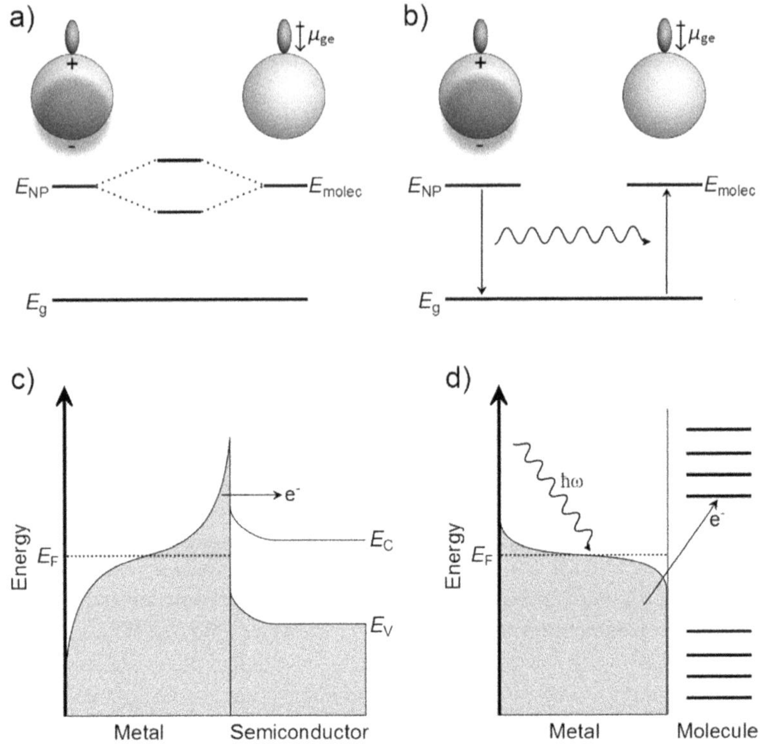

Figure 5. Schematic of major decay processes involving the interaction of the plasmonic metal with nearby molecules or semiconductors, including (a) hybridization of plasmonic and excitonic states, (b) plasmon resonance energy transfer, (c) hot electron transfer at a Schottky junction, and (d) direct absorption into charge-transfer states.

Photoinduced Electron Transfer

Energy absorbed by the plasmon can also be harnessed by transferring charge carriers, most commonly electrons, from the plasmonic system to nearby molecules or semiconductors. We note that charge transfer involving plasmonic nanoparticles may not necessarily result from excitation of the plasmon; many studies have looked at systems where the absorption of dye molecules leads to charge transfer from the excited dye molecule to the nanoparticle (*107–110*). Charge transfer resulting from excitation of the plasmonic nanoparticle may occur through several processes. The charge transfer may occur instantaneously by direct excitation into charge-transfer states, or may occur via excitation of the plasmon followed by decay to hot electrons, which then transfer to a nearby acceptor.

The prototypical example of hot electron transfer after plasmonic decay is in Au nanoparticles interacting with TiO_2 (*111–115*). The metal-semiconductor interface acts as a Schottky junction, where the metal Fermi level falls within the semiconductor band gap. If hot electrons in the metal are excited to energies above the Schottky barrier height, they can be injected into the semiconductor with an efficiency that depends on their energy (Figure 5c) (*116–118*). tunneling of lower-energy hot electrons across the barrier is also possible but occurs with much lower yield (*119*). Although hot-electron transfer is typically described for semiconductor acceptors, it has also been proposed as a mechanism for electron transfer to atomic (*120*) and molecular (*121*) acceptors.

The efficiency of hot-electron transfer in these structures depends on the population of hot electrons at energies above the Schottky barrier height, which is on the order of 1 eV between Au and TiO_2 (*122*). Although in many nanostructures the majority of hot electrons are typically excited to energies relatively close to the Fermi level (*30*, *31*, *49*, *55*), only electrons with higher energies contribute significantly to the electron transfer (*123*). Since smaller nanostructures tend to produce more hot electrons with high energy (*21*, *52*, *55*), theoretical modeling suggests that these nanostructures likewise have more hot carriers that cross the Schottky barrier (*55*). As mentioned earlier, the production of highly energetic carriers can be enhanced at hot spots (*45*); thus, it has been suggested that these hot spots may play an important role in increasing the rate of electron transfer (*124*).

Since hot electron thermalization is completed within a few hundreds of fs and the lifetimes of hot electrons with sufficient energy to cross the Schottky barrier is no more than a few tens of fs (*49*, *57*, *59*, *63*), hot-electron transfer also occurs on a very short time scale. Various experimental measurements suggest that the charge transfer occurs significantly faster than the thermalization timescale, with time estimates of less than 50 fs from Au nanoparticles to TiO_2 nanoparticles (*113*) and as short as 20 fs from Au nanoparticles to CdS nanorods (*125*). The limitation to high-energy carriers and the short time scale of thermalization limit the efficiency of the photoinduced electron transfer process, with most experimental and theoretical efficiency estimates less than 10% (*55*, *115*, *126*, *127*).

Hot-electron transfer is also dependent on the injection efficiency, defined as the probability that a hot electron at a particular energy will be transferred across the Schottky junction. The energy dependence of the transfer efficiency can be described in terms of a modified Fowler formula, which depends also on the device geometry and on the details of the interface (*116–118*). Since injection requires conservation of momentum between the initial and final states, increasing the amount of surface contact between the metal and the semiconductor can also improve the injection efficiency (*118*). If the interface is rough enough that the selection rules regarding conservation of momentum can be neglected, the theoretical injection rate can be greatly enhanced (*117*). The chemical details of the interface are also significant. In particular, TiO_2 oxygen vacancy sites at the interface with Au can bear negative charges and increase the effective height of the Schottky barrier; surface treatments to reduce these surface states can improve the injection efficiencies (*128*, *129*).

A second possible route to photoinduced electron transfer is direct electron transfer, where the plasmon-acceptor complex is excited directly into a charge-transfer state or the charge transfer is completed as part of the dephasing process of the plasmon (Figure 5d) (*130*). The dephasing-based mechanism is an extension of the chemical interface damping described before, which suggests transient transfer into adsorbate states as part of the dephasing process (*39–42*). This has been the primary electron transfer mechanism proposed experimentally and modeled for atomic (*131–133*) and molecular (*130*, *134–137*) acceptors, particularly for acceptors physisorbed or chemisorbed onto the metal surface. Experimental evidence for methylene blue on Ag nanocubes shows that the photoinduced electron transfer efficiency is strongly wavelength-dependent, which should not be the case if the plasmon decayed to a distribution of hot carriers before the electron transfer step (*130*). Theoretical models using a C-RT-TDDFT approach using Marcus theory to compute electron transfer rates between charge-localized diabatic states suggest that the quantum yield for electron transfer decays exponentially with increasing distance in the $[Ag_{20}\text{-}Ag]^+$ complex (*132*, *138*).

Although direct excitation has been primarily discussed in terms of molecular acceptors, recent evidence suggests that it may also play a significant role from plasmonic metals to semiconductors. Theoretical modeling of an Au_{20} cluster on a TiO_2 surface using ab initio molecular dynamics suggests that in this system, roughly half of charge carriers are generated instantaneously upon excitation and only half occur as a distinct hot-electron transfer step with a time scale of 40 fs (*139*). Since most experimental measures of the time scale for photoinduced electron transfer provide an upper limit rather than an exact estimate (*112*, *113*), it is possible that direct ET has played a larger role than has been typically recognized. Direct charge-transfer has also been recently proposed for Au tips on CdSe nanorods, as supported by very high quantum efficiencies of > 24% and the lack of the expected dependence of the efficiency on the excess energy as described by the Fowler equation (*140*).

Chemical Enhancement in SERS

Although the primary mechanism of SERS enhancement is due to plasmonic enhancement of the local electric fields (*6*, *86*, *87*, *141*), chemical interactions between the molecule and the nanoparticle can also contribute to the enhancement. Although the chemical enhancement is still somewhat controversial and not as clearly understood as the electromagnetic enhancement (*7*), the proposed mechanisms include resonance of molecular excitations with the incident light, the introduction of charge-transfer excited states between the metal and the adsorbate, and changes to the adsorbate ground state electronic structure due to interactions (such as ground state charge transfer) with the surface (*142*). Here, we focus on the latter two mechanisms.

Estimates of the magnitude of the chemical enhancement in SERS have varied widely. Early measurements of Raman scattering on smooth metal surfaces where the electromagnetic enhancement is known to be small estimated chemical enhancements on the order of $10\text{-}10^2$ (*143–145*). However, various approaches have suggested that the chemical enhancement may be more significant. The chemical enhancements in nanoparticle dimers have been proposed to be as large as 10^7 (*146*), and experiments on few-atom (*147*) and few-nm (*148*) clusters have shown enhancements larger than can be explained by electric field enhancement.

Quantum mechanical calculations suggest that the electric field may be smaller at close range (< 1 nm) than classical electrodynamics for individual nanoparticles (*149*) and dimers (*150*, *151*), which implies that the chemical enhancement must be larger than typically assumed based on the difference between experimental results and the classical electromagnetic enhancement.

One of the main routes of chemical enhancement is the introduction of charge-transfer states between the metal and the molecule, which is related to the charge-transfer involved in chemical interface damping as described earlier. This has long been proposed as a mechanism in SERS from experimental (*152–154*) and theoretical (*155*) perspectives, and electron energy loss spectroscopy suggests that surface adsorption introduces low-energy states related to charge transfer (*156*). Since the various enhancement mechanisms are challenging to separate experimentally, theoretical approaches can provide more insight. DFT-based models have shown charge-transfer enhancements on the order of 10^3 for the prototypical Ag_{20}-pyridine complex (*157*) and suggested that the metal-to-ligand charge-transfer states are of primary importance (*158*). Note that this work evaluated the SERS intensities in the static limit, so all excited states are virtual in evaluating the Raman intensity; however, other chemical enhancement theories assume that the charge transfer states are resonant, so there is a much stronger dependence on excited state energies and widths. The nature of the charge-transfer states is dependent on the system; for thiols on Au and Ag, consideration of the ligand-to-metal charge-transfer states is necessary to reproduce experimental trends (*159*). Since charge-transfer energies are well known to be greatly underestimated by most DFT functionals, the computed chemical enhancement is quite sensitive to the choice of functional (*160*).

Chemical interactions between the metal and the adsorbate in the ground state can also enhance the Raman signal, with experimental estimates of the enhancement on the order of $10\text{-}10^2$ (*161*, *162*). The magnitude of the chemical enhancement is dependent on the details of the molecule-metal interaction, including the identity of the metal (*163*, *164*), the adsorbate (*165*), the orientation dependence of the interaction (*166*), and intermolecular interactions between adsorbates (*167*). These interactions can also cause shifts of the Raman modes relative to the free molecules (*168*, *169*).

Conclusions and Outlook

Plasmonic enhancement of photochemical and spectroscopic processes is a promising route to a variety of applications. Although progress has been made toward solar energy conversion and sensing applications, significant challenges remain in understanding the underlying fundamental processes from both experimental and theoretical standpoints. Even though the full decay of the plasmons occurs via processes that take up to multiple ps, the relevant timescales for the electron-transfer-related processes critical for photochemical conversion and spectroscopic enhancement are typically no more than a few tens of fs. Much progress has been made toward experimentally resolving these processes on short time and length scales, but further work is needed to better resolve the time scales and nature of charge-transfer states. Theoretically, the size and complexity of typical plasmonic nanostructures has limited studies primarily to DFT-based quantum chemical modeling of small model nanoclusters and more approximate models of larger systems. These models have provided insight into the nature of the plasmonic states and their decay dynamics, but further work is needed to better understand the nature and role of the charge transfer states in chemical processes. Understanding of the fundamental photophysics may reveal important chemical features that can aid in designing better systems for plasmon-enhanced device and spectroscopic applications.

Acknowledgments

This research was supported by the Department of Energy grant DE-FG02-10ER16153 (for methods development), the NSF CaSTL Center grant CHE-1414466 (for time-resolved applications), and by AFOSR MURI grant FA9550-14-1-0003 (electrochemistry applications).

References

1. Clavero, C. *Nat. Photonics* **2014**, *8*, 95–103.
2. Brongersma, M. L.; Halas, N. J.; Nordlander, P. *Nat. Nanotechnol.* **2015**, *10*, 25–34.
3. Ma, X.-C.; Dai, Y.; Yu, L.; Huang, B.-B. *Light Sci. Appl.* **2016**, *5*, e16017.
4. Hou, W.; Cronin, S. B. *Adv. Funct. Mater.* **2013**, *23*, 1612–1619.
5. Jensen, L.; Aikens, C. M.; Schatz, G. C. *Chem. Soc. Rev.* **2008**, *37*, 1061–1073.
6. Willets, K. A.; Van Duyne, R. P. *Annu. Rev. Phys. Chem.* **2007**, *58*, 267–297.
7. Moskovits, M. *Phys. Chem. Chem. Phys.* **2013**, *15*, 5301–5311.
8. Mie, G. *Ann. Phys.* **1908**, *330*, 377–445.
9. Voshchinnikov, N. V.; Farafonov, V. G. *Astrophys. Space Sci.* **1993**, *204*, 19–86.
10. Yee, K. S. *IEEE Trans. Antennas Propag.* **1966**, *14*, 302–307.
11. Kelly, K. L.; Coronado, E.; Zhao, L. L.; Schatz, G. C. *J. Phys. Chem. B* **2003**, *107*, 668–677.
12. Burda, C.; Chen, X.; Narayanan, R.; El-Sayed, M. A. *Chem. Rev.* **2005**, *105*, 1025–1102.
13. Qin, L.; Zou, S.; Xue, C.; Atkinson, A.; Schatz, G. C.; Mirkin, C. a. *Proc. Natl. Acad. Sci. U.S.A.* **2006**, *103*, 13300–13303.
14. Camden, J. P.; Dieringer, J. A.; Wang, Y.; Masiello, D. J.; Marks, L. D.; Schatz, G. C.; Van Duyne, R. P. *J. Am. Chem. Soc.* **2008**, *130*, 12616–12617.
15. Kirschner, M. S.; Lethiec, C. M.; Lin, X.-M.; Schatz, G. C.; Chen, L. X.; Schaller, R. D. *ACS Photonics* **2016**, *3*, 758–763.
16. Gieseking, R. L.; Ratner, M. A.; Schatz, G. C. *J. Phys. Chem. A* **2016**, *120*, 4542–4549.
17. Guidez, E. B.; Aikens, C. M. *Nanoscale* **2014**, *6*, 11512–11527.

18. Aikens, C. M.; Li, S.; Schatz, G. C. *J. Phys. Chem. C* **2008**, *112*, 11272–11279.
19. Madison, L. R.; Ratner, M. A.; Schatz, G. C. In *Frontiers in Quantum Methods and Applications in Chemistry and Physics*; Nascimento, M. A. C., Ed.; Springer International Publishing: Switzerland, 2015; Vol. 29, pp 37–52.
20. Guidez, E. B.; Aikens, C. M. *Phys. Chem. Chem. Phys.* **2014**, *16*, 15501–15509.
21. Ma, J.; Wang, Z.; Wang, L.-W. *Nat. Commun.* **2015**, *6*, 10107.
22. López-Lozano, X.; Barron, H.; Mottet, C.; Weissker, H.-C. *Phys. Chem. Chem. Phys.* **2014**, *16*, 1820–1823.
23. van Dijk, M. A.; Tchebotareva, A. L.; Orrit, M.; Lippitz, M.; Berciaud, S.; Lasne, D.; Cognet, L.; Lounis, B. *Phys. Chem. Chem. Phys.* **2006**, *8*, 3486–3495.
24. Evanoff, D. D.; Chumanov, G. *J. Phys. Chem. B* **2004**, *108*, 13957–13962.
25. Langhammer, C.; Yuan, Z.; Zori, I.; Kasemo, B. *Nano Lett.* **2006**, *6*, 833–838.
26. Mooradian, A. *Phys. Rev. Lett.* **1969**, *22*, 185–187.
27. Dulkeith, E.; Niedereichholz, T.; Klar, T. A.; Feldmann, J.; Von Plessen, G.; Gittins, D. I.; Mayya, K. S.; Caruso, F. *Phys. Rev. B* **2004**, *70*, 1–4.
28. Mohamed, M. B.; Volkov, V.; Link, S.; El-Sayed, M. a. *Chem. Phys. Lett.* **2000**, *317*, 517–523.
29. Landau, L. *J. Phys.* **1946**, 25–34.
30. Govorov, A. O.; Zhang, H.; Demir, H. V.; Gun'Ko, Y. K. *Nano Today* **2014**, *9*, 85–101.
31. Marchuk, K.; Willets, K. A. *Chem. Phys.* **2014**, *445*, 95–104.
32. Sönnichsen, C.; Franzl, T.; Wilk, T.; von Plessen, G.; Feldmann, J.; Wilson, O.; Mulvaney, P. *Phys. Rev. Lett.* **2002**, *88*, 77402.
33. Liu, M.; Pelton, M.; Guyot-Sionnest, P. *Phys. Rev. B* **2009**, *79*, 35418.
34. Olmon, R. L.; Slovick, B.; Johnson, T. W.; Shelton, D.; Oh, S.-H.; Boreman, G. D.; Raschke, M. B. *Phys. Rev. B* **2012**, *86*, 235147.
35. Guo, Z.; Habenicht, B. F.; Liang, W. Z.; Prezhdo, O. V. *Phys. Rev. B* **2010**, *81*, 1–6.
36. Hubenthal, F. *Plasmonics* **2013**, *8*, 1341–1349.
37. Bosman, M.; Ye, E.; Tan, S. F.; Nijhuis, C. A.; Yang, J. K. W.; Marty, R.; Mlayah, A.; Arbouet, A.; Girard, C.; Han, M.-Y. *Sci. Rep.* **2013**, *3*, 1312.
38. Mittal, R.; Glenn, R.; Saytashev, I.; Lozovoy, V. V; Dantus, M. *J. Phys. Chem. Lett.* **2015**, *6*, 1638–1644.
39. Persson, B. N. J. *Surf. Sci.* **1993**, *281*, 153–162.
40. Pinchuk, A.; Kreibig, U. *New J. Phys.* **2003**, *5*.
41. Ziegler, T.; Hendrich, C.; Hubenthal, F.; Vartanyan, T.; Träger, F. *Chem. Phys. Lett.* **2004**, *386*, 319–324.
42. Bosbach, J.; Hendrich, C.; Stietz, F.; Vartanyan, T.; Träger, F. *Phys. Rev. Lett.* **2002**, *89*, 257404.
43. Hubenthal, F.; Träger, F. *Proc. SPIE* **2011**, *7922*, 79220D.

44. Almeida, E.; Moreira, A. C. L.; Brito-Silva, A. M.; Galembeck, A.; De Melo, C. P.; De S. Menezes, L.; De Araújo, C. B. *Appl. Phys. B* **2012**, *108*, 9–16.
45. Harutyunyan, H.; Martinson, A. B.; Rosenmann, D.; Khorashad, L. K.; Besteiro, L. V; Govorov, A. O.; Wiederrecht, G. P. *Nat. Nanotechnol.* **2015**, *10*, 770–774.
46. Tas, G.; Maris, H. J. *Phys. Rev. B* **1994**, *49*, 15046–15054.
47. Groeneveld, R. H. M.; Sprik, R.; Lagendijk, A. *Phys. Rev. B* **1992**, *45*, 5079–5082.
48. Link, S.; Burda, C.; Wang, Z. L.; El-Sayed, M. A. *J. Chem. Phys.* **1999**, *111*, 1255–1264.
49. Fann, W. S.; Storz, R.; Tom, H. W. K.; Bokor, J. *Phys. Rev. B* **1992**, *46*, 13592–13595.
50. Knoesel, E.; Hotzel, A.; Hertel, T.; Wolf, M.; Ertl, G. *Surf. Sci.* **1996**, *368*, 76–81.
51. Aeschlimann, M.; Bauer, M.; Pawlik, S.; Knorren, R.; Bouzerar, G.; Bennemann, K. H. *Appl. Phys. A* **2000**, *71*, 485–491.
52. Zhang, H.; Govorov, A. O. *J. Phys. Chem. C* **2014**, *118*, 7606–7614.
53. Manjavacas, A.; Nordlander, P.; Kulkarni, V.; Liu, J. G. *ACS Nano* **2014**, *8*, 7630–7638.
54. Sundararaman, R.; Narang, P.; Jermyn, A. S.; Goddard, W. a., III; Atwater, H. a. *Nat. Commun.* **2014**, *5*, 5788.
55. Govorov, A. O.; Zhang, H.; Gun'Ko, Y. K. *J. Phys. Chem. C* **2013**, *117*, 16616–16631.
56. Govorov, A. O.; Zhang, H. *J. Phys. Chem. C* **2015**, *119*, 6181–6194.
57. Bernardi, M.; Mustafa, J.; Neaton, J. B.; Louie, S. G. *Nat. Commun.* **2015**, *6*, 7044.
58. Ritchie, R. H.; Ashley, J. C. *J. Phys. Chem. Solids* **1965**, *26*, 1689–1694.
59. Del Fatti, N.; Voisin, C.; Achermann, M.; Tzortzakis, S.; Christofilos, D.; Vallée, F. *Phys. Rev. B* **2000**, *61*, 16956–16966.
60. Bauer, M.; Marienfeld, A.; Aeschlimann, M. *Prog. Surf. Sci.* **2015**, *90*, 319–376.
61. Merschdorf, M.; Kennerknecht, C.; Pfeiffer, W. *Phys. Rev. B* **2004**, *70*, 1–4.
62. Sun, C. K.; Vallée, F.; Acioli, L. H.; Ippen, E. P.; Fujimoto, J. G. *Phys. Rev. B* **1994**, *50*, 15337–15348.
63. Zhukov, V. P.; Aryasetiawan, F.; Chulkov, E. V.; Gurtubay, I. G. de; Echenique, P. M. *Phys. Rev. B* **2001**, *64*, 195122.
64. Masia, F.; Langbein, W.; Borri, P. *Phys. Rev. B* **2012**, *85*, 1–11.
65. Voisin, C.; Christofilos, D.; Del Fatti, N.; Vallée, F.; Prével, B.; Cottancin, E.; Lermé, J.; Pellarin, M.; Broyer, M. *Phys. Rev. Lett.* **2000**, *85*, 2200–2203.
66. Bauer, C.; Abid, J. P.; Fermin, D.; Girault, H. H. *J. Chem. Phys.* **2004**, *120*, 9302–9315.
67. Anisimov, S. I.; Kapeliovich, B. L.; Perel-man, T. L. *J. Exp. Theor. Phys.* **1974**, *66*, 375–377.
68. Ahmadi, T. S.; Logunov, S. L.; El-Sayed, M. A. *J. Phys. Chem.* **1996**, *100*, 8053–8056.

69. Aruda, K. O.; Tagliazucchi, M.; Sweeney, C. M.; Hannah, D. C.; Schatz, G. C.; Weiss, E. A. *Proc. Natl. Acad. Sci. U.S.A.* **2013**, *110*, 1–6.
70. Huang, J.; Wang, W.; Murphy, C. J.; Cahill, D. G. *Proc. Natl. Acad. Sci. U.S.A.* **2014**, *111*, 906–911.
71. Halté, V.; Bigot, J.; Palpant, B.; Broyer, M.; Prével, B.; Pérez, A. *Appl. Phys. Lett.* **2013**, *75*, 3799–3801.
72. Neukirch, A.; Guo, Z.; Prezhdo, O. V. *J. Phys. Chem. C* **2012**, *116*, 15034–15040.
73. Ranasingha, O.; Wang, H.; Zobač, V.; Jelinek, P.; Panapitiya, G.; Neukirch, A. J.; Prezhdo, O. V; Lewis, J. P. *J. Phys. Chem. Lett.* **2016**, *7*, 1563–1569.
74. Groeneveld, R. H. M.; Sprik, R.; Lagendijk, A. *Phys. Rev. B* **1994**, *51*, 11433.
75. Logunov, S. L.; Ahmadi, T. S.; El-Sayed, M. A.; Khoury, J. T.; Whetten, R. L. *J. Phys. Chem. B* **1997**, *101*, 3713–3719.
76. Link, S.; Hathcock, D. J.; Nikoobakht, B.; El-Sayed, M. A. *Adv. Mater.* **2003**, *15*, 393–396.
77. Westcott, S. L.; Averitt, R. D.; Wolfgang, J. A.; Nordlander, P.; Halas, N. J. *J. Phys. Chem. B* **2001**, *105*, 9913–9917.
78. Hu, M.; Hartland, G. V. *J. Phys. Chem. B* **2002**, *106*, 7029–7033.
79. Baffou, G.; Quidant, R. *Laser Photon. Rev.* **2013**, *7*, 171–187.
80. Mubeen, S.; Lee, J.; Lee, W.; Singh, N.; Stucky, G. D.; Moskovits, M. *ACS Nano* **2014**, *8*, 6066–6073.
81. Atwater, H. A.; Polman, A. *Nat. Mater.* **2010**, *9*, 865–865.
82. Awazu, K.; Fujimaki, M.; Rockstuhl, C.; Tominaga, J.; Murakami, H.; Ohki, Y.; Yoshida, N.; Watanabe, T. *J. Am. Chem. Soc.* **2008**, *130*, 1676–1680.
83. Linic, S.; Christopher, P.; Xin, H.; Marimuthu, A. *Acc. Chem. Res.* **2013**, *46*, 1890–1899.
84. Eustis, S.; El-Sayed, M. A. *Chem. Soc. Rev.* **2006**, *35*, 209–217.
85. Thomas, K. G.; Kamat, P. V. *Acc. Chem. Res.* **2003**, *36*, 888–898.
86. Nie, S.; Emory, S. R. *Science* **1997**, *275*, 1102–1107.
87. Kneipp, K.; Wang, Y.; Kneipp, H.; Perelman, L. T.; Itzkan, I.; Dasari, R. R.; Feld, M. S. *Phys. Rev. Lett.* **1997**, *78*, 1667–1670.
88. Chen, H.; Schatz, G. C.; Ratner, M. A. *Rep. Prog. Phys.* **2012**, *75*, 96402.
89. Fofang, N. T.; Park, T. H.; Neumann, O.; Mirin, N. A.; Nordlander, P.; Halas, N. J. *Nano Lett.* **2008**, *8*, 3481–3487.
90. Hao, Y. W.; Wang, H. Y.; Jiang, Y.; Chen, Q. D.; Ueno, K.; Wang, W. Q.; Misawa, H.; Sun, H. B. *Angew. Chem. Int. Ed. Engl.* **2011**, *50*, 7824–7828.
91. Eizner, E.; Avayu, O.; Ditcovski, R.; Ellenbogen, T. *Nano Lett.* **2015**, *15*, 6215–6221.
92. Manjavacas, A.; Abajo, F. J. G. De; Nordlander, P. *Nano Lett.* **2011**, *11*, 2318–2323.
93. Nascimento, D. R.; DePrince, A. E., III *J. Chem. Phys.* **2015**, *143*, 214104.
94. Fihey, A.; Maurel, F.; Perrier, A. *J. Phys. Chem. C* **2015**, *119*, 9995–10006.
95. Marinica, D. C.; Lourenço-Martins, H.; Aizpurua, J.; Borisov, A. G. *Nano Lett.* **2013**, *13*, 5972–5978.

96. Sukharev, M.; Seideman, T.; Gordon, R. J.; Salomon, A.; Prior, Y. *ACS Nano* **2014**, *36*, 807–817.
97. Liu, G. L.; Long, Y.-T.; Choi, Y.; Kang, T.; Lee, L. P. *Nat. Methods* **2007**, *4*, 1015–1017.
98. Jimenez, J. A.; Lysenko, S.; Liu, H. *J. Appl. Phys.* **2008**, *104*, 054313.
99. Choi, Y.; Kang, T.; Lee, L. P. *Nano Lett.* **2009**, *9*, 85–90.
100. Mol, B.; Joy, L. K.; Thomas, H.; Thomas, V.; Joseph, C.; Narayanan, T. N.; Al-Harthi, S.; Unnikrishnan, N. V; Anantharaman, M. R. *Nanotechnology* **2016**, *27*, 85701.
101. Li, J.; Cushing, S. K.; Meng, F.; Senty, T. R.; Bristow, A. D.; Wu, N. *Nat. Photonics* **2015**, *9*, 601–607.
102. Cushing, S. K.; Li, J.; Bright, J.; Yost, B. T.; Zheng, P.; Bristow, A. D.; Wu, N. *J. Phys. Chem. C* **2015**, *119*, 16239–16244.
103. Wang, M.; Bangalore Rajeeva, B.; Scarabelli, L.; Perillo, E. P.; Dunn, A.; Liz-Marzán, L. M.; Zheng, Y. *J. Phys. Chem. C* **2016**, *120*, 14820–14827.
104. Jimenez, J. A.; Lysenko, S.; Vikhnin, V. S.; Liu, H. *J. Electron. Mater.* **2010**, *39*, 138–143.
105. Razgoniaeva, N.; Lambright, S.; Sharma, N.; Acharya, A.; Khon, E.; Moroz, P.; Razgoniaev, A.; Ostrowski, A.; Zamkov, M. *J. Phys. Chem. C* **2015**, *119*, 15562–15571.
106. Cushing, S. K.; Li, J.; Meng, F.; Senty, T. R.; Suri, S.; Zhi, M.; Li, M.; Bristow, A. D.; Wu, N. *J. Am. Chem. Soc.* **2012**, *134*, 15033–15041.
107. Ipe, B. I.; Thomas, K. G.; Barazzouk, S.; Hotchandani, S.; Kamat, P. V. *J. Phys. Chem. B* **2002**, *106*, 18–21.
108. Kotiaho, A.; Lahtinen, R. M.; Tkachenko, N. V; Efimov, A.; Kira, A.; Imahori, H.; Lemmetyinen, H. *Langmuir* **2007**, *23*, 13117–13125.
109. Kotiaho, A.; Lahtinen, R.; Efimov, A.; Metsberg, H.-K.; Sariola, E.; Lehtivuori, H.; Tkachenko, N. V; Lemmetyinen, H. *J. Phys. Chem. C* **2010**, *114*, 162–168.
110. Pramod, P.; Soumya, C. C.; Thomas, K. G. *J. Phys. Chem. Lett.* **2011**, *2*, 775–781.
111. Zhao, G.; Kozuka, H.; Yoko, T. *Thin Solid Films* **1996**, *277*, 147–154.
112. Furube, A.; Du, L.; Hara, K.; Katoh, R.; Tachiya, M. *J. Am. Chem. Soc.* **2007**, *129*, 14852–14853.
113. Du, L.; Furube, A.; Yamamoto, K.; Hara, K.; Katoh, R.; Tachiya, M. *J. Phys. Chem. C* **2009**, 6454–6462.
114. Mubeen, S.; Hernandez-Sosa, G.; Moses, D.; Lee, J.; Moskovits, M. *Nano Lett.* **2011**, *11*, 5548–5552.
115. Mubeen, S.; Lee, J.; Singh, N.; Krämer, S.; Stucky, G. D.; Moskovits, M.; Kramer, S.; Stucky, G. D.; Moskovits, M. *Nat. Nanotechnol.* **2013**, *8*, 247–251.
116. Fowler, R. H. *Phys. Rev.* **1931**, *38*, 107.
117. Giugni, a.; Torre, B.; Toma, A.; Francardi, M.; Malerba, M.; Alabastri, A.; Proietti Zaccaria, R.; Stockman, M. I.; Di Fabrizio, E. *Nat. Nanotechnol.* **2013**, *8*, 845–852.
118. Knight, M. W.; Wang, Y.; Urban, A. S.; Sobhani, A.; Zheng, B. Y.; Nordlander, P.; Halas, N. J. *Nano Lett.* **2013**, *13*, 1687–1692.

119. Moskovits, M. *Science.* **2011**, *332*, 676–677.
120. Kim, K.; Lee, S. H.; Choi, J.-Y.; Shin, K. S. *J. Phys. Chem. C* **2014**, *118*, 3359–3365.
121. Conklin, D.; Nanayakkara, S.; Park, T.-H.; Lagadec, M. F.; Stecher, J. T.; Chen, X.; Therien, M. J.; Bonnell, D. A. *ACS Nano* **2013**, *7*, 4479–4486.
122. McFarland, E. W.; Tang, J. *Nature* **2003**, *421*, 616–618.
123. Kumarasinghe, C. S.; Premaratne, M.; Bao, Q.; Agrawal, G. P. *Sci. Rep.* **2015**, *5*, 12140.
124. Linic, S.; Christopher, P.; Ingram, D. B. *Nat. Mater.* **2011**, *10*, 911–921.
125. Mongin, D.; Shaviv, E.; Maioli, P.; Crut, A.; Banin, U.; Fatti, N. Del; Vallée, F. *ACS Nano* **2012**, *6*, 7034–7043.
126. Wu, K.; Rodríguez-Córdoba, W. W. E.; Yang, Y.; Lian, T.; Rodr, W. E.; Yang, Y.; Lian, T.; Rodríguez-Córdoba, W. W. E.; Yang, Y.; Lian, T. *Nano Lett.* **2013**, *13*, 5255–5263.
127. Hoggard, A.; Wang, L.; Ma, L.; Fang, Y.; You, G.; Olson, J.; Liu, Z. *ACS Nano* **2013**, *7*, 11209–11217.
128. Tamura, T.; Ishibashi, S.; Terakura, K.; Weng, H. *Phys. Rev. B* **2009**, *80*, 1–6.
129. Ma, X.; Dai, Y.; Yu, L.; Huang, B. *ACS Appl. Mater. Interfaces* **2014**, *6*, 12388–12394.
130. Boerigter, C.; Campana, R.; Morabito, M.; Linic, S. *Nat. Commun.* **2016**, *7*, 10545.
131. Petek, H. *Science* **2000**, *288*, 1402–1404.
132. Chen, H.; Ratner, M. A.; Schatz, G. C. *J. Phys. Chem. C* **2011**, *115*, 18810–18821.
133. Chen, H.; Ratner, M. A.; Schatz, G. C. *J. Phys. Chem. C* **2014**, *118*, 1755–1762.
134. Jain, P. K.; Qian, W.; El-Sayed, M. A. *J. Am. Chem. Soc.* **2006**, *128*, 2426–2433.
135. Kale, M. J.; Kale, M. J.; Avanesian, T.; Christopher, P. *ACS Catal.* **2014**, *4*, 116–128.
136. Robatjazi, H.; Bahauddin, S. M.; Doiron, C.; Thomann, I. *Nano Lett.* **2015**, *15*, 6155–6161.
137. Boerigter, C.; Aslam, U.; Linic, S. *ACS Nano* **2016**acsnano.6b01846.
138. Chen, H.; Ratner, M. A.; Schatz, G. C. *J. Photochem. Photobiol. A* **2011**, *221*, 143–147.
139. Long, R.; Prezhdo, O. V. *J. Am. Chem. Soc.* **2014**, *136*, 4343–4354.
140. Wu, K.; Chen, J.; Mcbride, J. R.; Lian, T. *Science* **2015**, *349*, 632–635.
141. Moskovits, M. *J. Raman Spectrosc.* **2005**, *36*, 485–496.
142. Morton, S. M.; Silverstein, D. W.; Jensen, L. *Chem. Rev.* **2011**, *111*, 3962–3994.
143. Udagawa, M.; Chou, C. C.; Hemminger, J. C.; Ushioda, S. *Phys. Rev. B* **1981**, *23*, 6843–6846.
144. Campion, A.; Mullins, D. R. *Chem. Phys. Lett.* **1983**, *94*, 576–579.
145. Jiang, X.; Campion, A. *Chem. Phys. Lett.* **1987**, *140*, 95–100.
146. Fromm, D. P.; Sundaramurthy, A.; Kinkhabwala, A.; Schuck, P. J.; Kino, G. S.; Moerner, W. E. *J. Chem. Phys.* **2006**, *124*, 1–4.

147. Peyser-Capadona, L.; Zheng, J.; González, J.; Lee, T.-H.; Patel, S.; Dickson, R. *Phys. Rev. Lett.* **2005**, *94*, 58301.
148. Zheng, J.; Ding, Y.; Tian, B.; Zhong, L. W.; Zhuang, X. *J. Am. Chem. Soc.* **2008**, *130*, 10472–10473.
149. Zuloaga, J.; Prodan, E.; Nordlander, P. *ACS Nano* **2010**, *4*, 5269–5276.
150. Zuloaga, J.; Prodan, E.; Nordlander, P. *Nano Lett.* **2009**, *9*, 887–891.
151. Esteban, R.; Borisov, A. G.; Nordlander, P.; Aizpurua, J. *Nat. Commun.* **2012**, *3*, 825.
152. Otto, A. *Surf. Sci.* **1980**, *92*, 145–152.
153. Otto, A.; Timper, J.; Billmann, J.; Kovacs, G.; Pockrand, I. *Surf. Sci.* **1980**, *92*, 55–57.
154. Otto, A. *J. Electron. Spectrosc. Relat. Phenom.* **1983**, *29*, 329–342.
155. Persson, B. N. J. *Chem. Phys. Lett.* **1981**, *82*, 561–565.
156. Campion, A.; Ivanecky, J. E., III; Child, C. M.; Foster, M. *J. Am. Chem. Soc.* **1995**, *117*, 11807–11808.
157. Zhao, L. L.; Jensen, L.; Schatz, G. C. *J. Am. Chem. Soc.* **2006**, *128*, 2911–2919.
158. Morton, S. M.; Jensen, L. *J. Am. Chem. Soc.* **2009**, *131*, 4090–4098.
159. Valley, N.; Greeneltch, N.; Van Duyne, R. P.; Schatz, G. C. *J. Phys. Chem. Lett.* **2013**, *4*, 2599–2604.
160. Moore, J. E.; Morton, S. M.; Jensen, L. *J. Phys. Chem. Lett.* **2012**, *3*, 2470–2475.
161. Kim, N.-J.; Lin, M.; Hu, Z.; Li, H. *Chem. Commun. (Camb).* **2009** (41), 6246–6248.
162. Maitani, M. M.; Ohlberg, D. A. A.; Li, Z.; Allara, D. L.; Stewart, D. R.; Williams, R. S. *J. Am. Chem. Soc.* **2009**, *131*, 6310–6311.
163. Wu, D. Y.; Ren, B.; Tian, Z. Q. *Isr. J. Chem.* **2006**, *46*, 317–327.
164. Wu, D.-Y.; Liu, X.-M.; Duan, S.; Xu, X.; Ren, B.; Lin, S.-H.; Tian, Z.-Q. *J. Phys. Chem. C* **2008**, *112*, 4195–4204.
165. Morton, S. M.; Ewusi-Annan, E.; Jensen, L. *Phys. Chem. Chem. Phys.* **2009**, *11*, 7348.
166. Saikin, S. K.; Olivares-Amaya, R.; Rappoport, D.; Stopa, M.; Aspuru-Guzik, A. *Phys. Chem. Chem. Phys.* **2009**, *11*, 9401–9411.
167. Cabalo, J.; Guicheteau, J. A.; Christesen, S. *J. Phys. Chem. A* **2013**, *117*, 9028–9038.
168. Wu, D. Y.; Ren, B.; Jiang, Y. X.; Xu, X.; Tian, Z. Q. *J. Phys. Chem. A* **2002**, *106*, 9042–9052.
169. Saikin, S. K.; Chu, Y.; Rappoport, D.; Crozier, K. B.; Aspuru-Guzik, A. *J. Phys. Chem. Lett.* **2010**, *1*, 2740–2746.

Chapter 2

Near-Field Interaction between Single Molecule and an Electromagnetic Field at "Hotspot" Generated by Plasmon Resonance

Tamitake Itoh*,[1] and Yuko S. Yamamoto[2,3]

[1]Health Technology Research Center,
National Institute of Advanced Industrial Science and Technology (AIST),
Takamatsu, Kagawa 761-0395, Japan
[2]Research Fellow of the Japan Society for the Promotion of Science,
Chiyoda, Tokyo 102-8472, Japan
[3]Department of Advanced Materials Science, Faculty of Engineering,
Kagawa University, Takamatsu, Kagawa 761-0396, Japan
*E-mail: tamitake-itou@aist.go.jp

Electromagnetic (EM) coupling between plasmon and molecular dipoles within optical near-field (NF) at metallic nanoparticle (NP) junctions, namely, "hotspots", enhances optical responses of the molecule. The enhancement enables even single molecule (SM) Raman detection. This enhancement phenomenon has been explained by the EM mechanism, in which the enhancement is described as temporal and spatial confinement effect of EM fields by plasmon resonance. In this chapter, we first focus on explanation of the EM mechanism, second discuss the limitation of the EM mechanism from the points of EM field intensity and EM field confinement, and final show the frontiers beyond the limitation of the EM mechanism e.g., ultrafast dynamics, strong coupling, and the field-gradient effect using the breakdowns of Fermi's golden rules. We then conclude this topic by showing summary and outlook.

Introduction

NF interaction between plasmon and an EM field of light at metallic NP enables to confine the EM field within several tens nanometers in the vicinity of the NP surface (*1*). If such NPs form aggregates like dimers, the EM field is further confined within one nanometer in the gap of the NP dimer (*2, 3*). Thus, the light intensity in such gap is extremely high compared with the light intensity in a free space, and the gap is called "hotspot". Various new applications e.g. ultra-sensitive molecular spectroscopy and unique photochemical reaction have been widely investigated at the hotspots in such NP aggregates (*4*). However, the framework of the EM coupling between the confined EM field and molecules at hotspots might be different from the theoretical one used for conventional spectroscopy, because conditions of such EM coupling are deviated from assumptions in the theory of the conventional spectroscopy in some respects (*5*).

In the book chapter, first we explain the framework of EM coupling at hotspots using the theoretical framework of conventional spectroscopy; second we estimate the applicable limitations of the framework used for conventional spectroscopy from the points of view of EM field intensity and EM field confinement, and the last we introduce interesting phenomena occurring when EM coupling beyond the applicable limitations of the framework. We use the EM mechanism of surface-enhanced Raman scattering (SERS), which is the most common molecular spectroscopy at hotspots, for the explanations of a classical theoretical framework of EM coupling. We also use SERS to explain the applicable limitations of the classical framework. As the interesting phenomena at hotspots, we introduce three phenomena: ultra-fast surface enhanced fluorescence (ultra-fast SEF), strong coupling between plasmon and single molecule, and allowance of forbidden transitions. For the sake of simplicity, we use rhodamine 6G (R6G) as the target molecule to probing the NF interaction at hotspots.

The Electromagnetic Mechanism of SERS

We here explain the EM coupling at hotspots using the classical framework used for explanation of SERS. The framework is called the EM mechanism, in which plasmon resonance of metal nanostructure e.g., NPs enhances the Raman cross-section of molecules on the NPs (*6*). Plasmon is collective oscillation of conduction electrons, which induce depolarization to cancel the external electric field of light (*7*). When the phase of the depolarization cannot follows the phase of the external field and their phase difference between the depolarization and external field reaches $\pi/2$, they become resonant. Due to the resonance, the external electric fields is temporally and spatially confined around the NPs. Specific metals, which have rich conduction electrons and whose plasmon resonance energies are deviated from their interband transition energies, can exhibit strong plasmon resonance. The temporal and spatial confinement results in enhancement of optical transition probability of a molecule on the NPs. The factor of the enhancement is expressed by Purcell factor $F \sim (\lambda/n)^3Q/V$, where λ is wavelength of light (angular frequency of light $\omega = (c/\lambda)2\pi$, where c is velocity of light in a vacuum), n is a refractive index of the hotspot, Q is a quality factor of the resonance, and V is mode volume of light at the hotspot (*2*). We here assume a rigorous resonant condition that plasmon resonance angular frequency $\omega_p = \omega_{ex}$. The value of Q factor is derived by dividing the resonance angular frequency ω_p with resonance linewidth $\Delta\omega_p$ and means how many times the confined electric field oscillates at a hotspot (how many times the electric field interacts with the molecule) compared with a free space. Thus the space with a higher Q factor can more efficiently excite a molecule and can also efficiently de-excite the excited molecule than a free space. In other words, a Q factor expresses temporal enhancement of interaction probability between the light and the molecule. The value of a Q factor of plasmon resonance at a hotspot of a dimer is around ten times. The value of $(\lambda/n)^3/V$ means how many times the electric field is confined at a hotspot V compared with the electric field of a free space $(\lambda/n)^3$. Thus the hotspot with a higher $(\lambda/n)^3/V$ can more efficiently excite the molecule and can also efficiently de-excite the excited molecule than a free space. In other words, $(\lambda/n)^3/V$ expresses spatial enhancement of interaction probability between light and a molecule. The value of $(\lambda/n)^3/V$ at a hotspot of a dimer, which exhibits a maximum value of $(\lambda/n)^3/V$ and enables SM SERS (*8–10*), is around 10^4 times, indicating V ~several nm^3. Thus, Purcell factor at a hotspot of a dimer becomes around 10^5 (10×10^4) times (the value is sometimes reported ~10^7 times) (*2, 3*).

Kerker et al applied Purcell factor of plasmon resonance to explain SERS (*1*). A Raman process is composed of an excitation transition and a Raman emission transition. Thus, both transitions are enhanced by the factor of F, and total enhancement factor of Raman cross-section becomes 10^{10} $(10^5\times10^5)$ times. In the case of resonant Raman process, whose cross-section is around ~10^{-24} cm^2, SERS cross-section becomes 10^{-14} cm^2 $(10^{10}\times10^{-24}$ $cm^2)$. This value is 10^2 times higher than a fluorescence cross-section of single molecule, and single molecule SERS observation is realized under low excitation power ~several W/cm^2 (*11*).

We explain the formulation and experimental evaluation of SERS using the EM mechanism. In fact, $\omega_p = \omega_{ex}$ is not collect because of Stokes shift of Raman photon energy. Thus, we use the enhancement factor of the electric field amplitude $|M| = |E_{loc}|/|E_I|$, where E_I and E_{loc} indicate the amplitudes of the incident and enhanced local electric field, respectively, instead of Purcell factor ($F = |M|^2$, when $\omega_p = \omega_{ex}$). Thanks to the broad linewidth of plasmon resonance $\Delta\omega_p\hbar$ ~100 meV (corresponding to the linewidth of ~50 nm at 600 nm), the resonance can enhance both the Raman excitation and Raman emission probabilities. The Raman excitation (emission) enhancement factor is described as a ratio of EM field intensities with and without a decrease in the mode volume of states of light by plasmon resonance. Thus, we have (*1–3*)

$$|M(\lambda_L,d_{av})|^2\,|M(\lambda_{em},d_{av})|^2 = |E_{loc}(\lambda_L,d_{av})|^2/|E_I(\lambda_L)|^2\,|E_{loc}(\lambda_{em},d_{av})|^2/|E_I(\lambda_{em})|^2, \quad (1)$$

where λ_L and λ_{em} denote the wavelengths of the incident and emission light, respectively, and d_{av} is the "effective" distance between a molecule and metal surface. Here, the term "effective" means that d_{av} includes both separation and orientation of the molecule. Thus, the SERS cross section $\sigma_{SERS}(\lambda_L, \lambda_{em})$ is given by

$$\sigma_{SERS}(\lambda_L, \lambda_{em}, d_{av}) = |M(\lambda_L, d_{av})|^2 |M(\lambda_{em}, d_{av})|^2 \sigma_{RS}(\lambda_L, \lambda_{em}), \quad (2)$$

where $\sigma_{RS}(\lambda_L, \lambda_{em})$ is Raman scattering cross-section of a molecule in free space. The EM mechanism has been experimentally investigated by many researchers.

The experimental evaluation of the EM mechanism using silver NP dimers dominantly showing dipolar plasmon resonance and a rhodamine 6G (R6G) molecule located at a hotspot of the dimer are explained. The detection of dipolar plasmons by dark-field microspectroscopy is important for reducing complexity in the investigation e.g., various strong coupling among plasmon modes in NP aggregates (*12*). First, the EM coupling of a plasmon and a molecular dipole was observed by measuring a common polarization dependence of the plasmon resonance spectra and SERS spectra from identical silver NP dimers having a R6G molecule at the hotspot (*13*). This observation is important for identifying the plasmon involved with the EM enhancement. Second, the Raman enhancement by plasmon resonance was confirmed by showing the relationship between SERS intensities and Q factors of plasmon resonances, because EM enhancement factor is identical to Purcell factor (*14*). This confirmation is important because nonradiative plasmons may also work as de-enhancement of Raman intensity when the rate of energy transfer from a resonantly excited molecular dipole to nonradiative plasmon modes becomes faster than dephasing rate of the molecular dipoles. A theoretic work indicates that such situation is realistic for resonance Raman processes of molecule very close to a metal surface (*15*). In other words, the confirmation showed that the "enhancement" is indeed induced by the plasmon resonance. Third, the excitation enhancement factor $|M(\lambda_L)|^2$ in Eq. (1) was evaluated using the excitation laser energy dependence of the SERS intensities (*16*). Fourth, the emission enhancement factor $|M(\lambda_{em})|^2$ in Eq. (1) was evaluated as the plasmon-induced spectral modulation of the SERS intensities from the points of dimer-by-dimer variations in plasmon resonance and surrounding medium-by-medium variations in refractive indexes (*17–19*). Finally, the experimental values for $\sigma_{SERS}(\lambda_L,\lambda_{em})$ are quantitatively reproduced by the product of $|M(\lambda_L)|^2|M(\lambda_{em})|^2$ and $\sigma_{RS}(\lambda_L,\lambda_{em})$ as in Eq. (2). $|M(\lambda_L)|^2|M(\lambda_{em})|^2$s were calculated with a finite-difference time-domain (FDTD) method using silver NP dimers and a R6G molecule located in the hotspots (*11*). This reproduction of experimentally-obtained SERS spectra by calculation results of $|M(\lambda_L)|^2|M(\lambda_{em})|^2$ and $\sigma_{RS}(\lambda_L,\lambda_{em})$ demonstrates that the EM mechanism is dominant in SERS phenomena. Figure 1 shows the excitation laser energy dependence of the reproduced SERS spectra in Ref. (*18*). The strange SERS spectral properties e.g., combination and overtone line intensities comparable to fundamental line intensities and anti-Stokes intensities higher than Stokes intensities, are quantitatively reproduced by the EM mechanism.

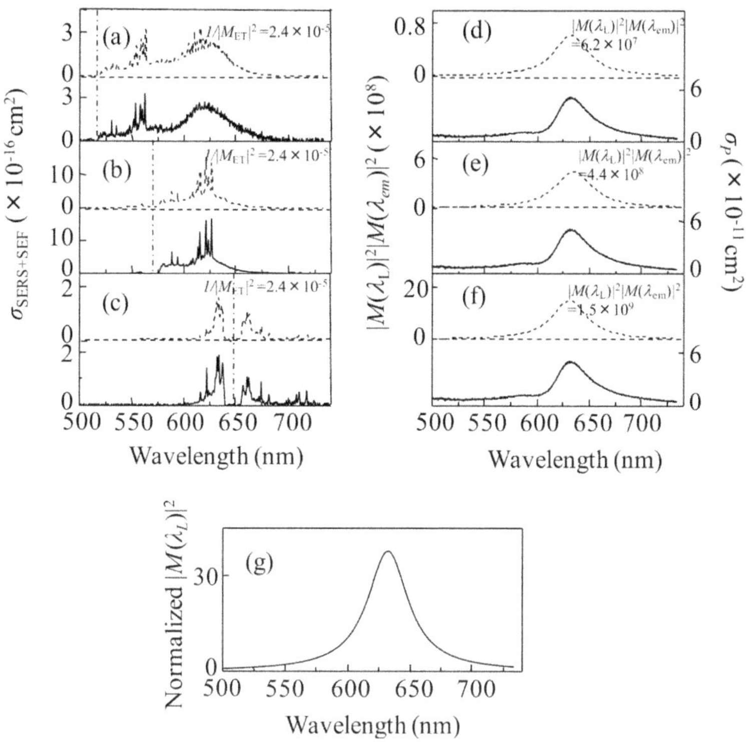

Figure 1. (a)-(c) SERS and SEF spectra calculated by the EM mechanism (dashed lines) and experimental SERS and SEF spectra (solid lines) excited at 514, 568, and 647 nm of R6G. Note that $|M_{ET}|^2$ means an enhancement factor of a quenching rate q of fluorescence. (d)-(f) Spectra of $|M(\lambda_L)|^2|M(\lambda_{em})|^2$ (dashed lines) and experimental plasmon resonance spectra (solid lines). Note that left and right axes show $|M(\lambda_L)|^2|M(\lambda_{em})|^2$ and the cross section of experimental plasmon resonance spectra, respectively. (g) Wavelength dependence of $|M(\lambda_L)|^2$. The values of $|M(\lambda_L)|^2$ are normalized at 514 nm. The values of q are inserted in (a)-(c); and the maximum values of $|M(\lambda_L)|^2|M(\lambda_{em})|^2$ are inserted in (d)-(f). Reproduced from Ref. (18). Copyright 2009, American Physical Society.

Key of this successful reproduction is that a dipolar plasmon reserves and temporally and spatially concentrates the excitation light at a hotspot, where the excitation light changes into Raman light, and then radiates the Raman light from the hotspot by inverse process of excitation. We also investigated that The EM mechanism is dominant in surface enhanced hyper Raman scattering (SEHRS) and surface enhanced fluorescence (SEF) (*20–22*).

Nonlinear Optical Response of Molecules and Unscreened Surface Electrons of Metal Determining the Limitation of the Electromagnetic Mechanism

We explained that the values of $|M(\lambda_L)|^2|M(\lambda_{em})|^2$s up to 10^{10} at hotspots enable SM SERS detections with excitation intensity ~several cm²/W without destructing the molecule (*11*). However, there are limitations of the EM mechanism to evaluate optical responses of molecules at hotspots. Indeed, if the electric field intensity exceeds a certain value, molecules cannot linearly response to the field. If the physical volume of a hotspot becomes smaller than a certain value, the value of V confined by plasmon resonance may not be the same as the physical volume. Several experiments and calculations have implicitly or explicitly indicated such limitations (*5*). Here these limitations are explained by changing the parameters describing the EM mechanism. First we explain the limitation of the linearly response by increasing incident light intensity $I_{in} = (1/2)c\varepsilon_0 n|E_I|^2$ regarding vibrational pumping effect and Rabi oscillation effect, where c and ε_0 are velocity of light in vacuum 3.0×10^8 m/s and permittivity of vacuum ~8.85×10^{-12} F/m. Second we explain the limitation of the value of V appearing as the effect of unscreened surface electron gas.

When one increases incident light intensity, vibrational pumping effect firstly appears as the limitation of the EM mechanism. Vibrational pumping effect means excitation of molecules in the vibrationally excited state in an electrically ground state. Thus, the ratio between Stokes and anti-Stokes SERS lines deviates from the ratio expected by the EM mechanism due to a quadratical increase in anti-Stokes SERS intensity against I_{in} even Stokes SERS intensity linearly increases against I_{in}. Such vibrational pumping effect in SERS process was firstly observed by K. Kneipp et al (*23*). We estimate the value of I_{in} which makes the effect important. The effect becomes important when the excitation rate becomes larger than vibrational damping rates. Xu et al estimated the value of I_{in} using a R6G molecule located at a hotspot (*24*). The excitation rate for a molecule located at the hotspot is roughly estimated to be $|M|^2\sigma_A\varphi_{ex}$, where σ_A is an absorption cross-section (not σ_{RS}) and φ_{ex} is photon flux. Using typical vibrational damping rate $\gamma_{vib} \sim 10^{12}$ s^{-1}, $|M|^2 \sim 10^5$ at hotspots, and $\sigma_A \sim 10^{-16}$ cm² for a R6G molecule at the excitation laser energy $\hbar\omega_L = 2.33$ eV (532 nm), where $\hbar$ is reduced Planck constant 1.06×10^{-34} Js, $\varphi_{ex} \sim 10^{23}$ photons/(cm²s) corresponding to I_{in} ~50 kW/cm² makes the value of $|M|^2\sigma_A\varphi_{ex}$ comparable to 10^{12} s^{-1}, resulting in anomaly large anti-Stokes SERS intensity compared with Stokes SERS intensity. For such I_{in}, Stokes/anti-Stokes ratios of SERS spectral shapes are deviated from those predicted the EM mechanism by Eq. (2).

When one increases incident light intensity, Rabi oscillation effect secondly appears as the limitation of the EM mechanism. When a molecule is excited by a coherent laser beam, the molecule cyclically absorbs photons and re-emits them as stimulated emission (*25*). The frequency of this cycle is called Rabi frequency and the cycle is called Rabi oscillation. Using the two level system, the Rabi frequency under the rigorous resonant condition can be derived as $\Omega_R = |M|p_0E_I/\hbar$, where p_0 is a transition dipole moment of a R6G molecule and E_I is incident electric-field amplitude (*26*). If Ω_R becomes comparable to the dephasing rate of the electronic transition dipole $\gamma_{ph} \sim 10^{14}$ s^{-1}, Rabi oscillation cannot be neglected, resulting in decrease in SERS intensity by the portion of the stimulated emission. Xu et al. estimated the value of I_{in} realizing such effect. The $p_0 = el_{dip}$, where e is elementary charge of an electron ~1.602×10^{-19} C and l_{dip} is a dipole length ~0.1 nm for a R6G molecule. Using $|M|^2$ ~10^5 at hotspots, $E_I = 1.5 \times 10^5$ V/m, which corresponds to I_{in} of ~600 kW/cm^2, makes $\Omega_R \sim 10^{14}$ s^{-1}. The value of Ω_R is comparable to $\gamma_{ph} \sim 10^{14}$ s^{-1}, resulting in decrease in SERS intensity even increasing in I_{in}. For such I_{in}, SERS intensities are largely deviated from those predicted by Eq. (2). The estimated values of I_{in} showing such limitations of EM mechanism are in the range from 50 to 600 kW/cm^2. The range is much higher than standard excitation conditions for SM SERS experiments ~several W/cm^2 (*11*).

When one decreases the mode volume of light confined by plasmon resonance at the hotspot, the effect of unscreened surface electron gas should be considered as the limitation of the EM mechanism (*27*). The limitation of V is determined by the volume of such electron gas within the Debye radius <0.5 nm from metal surfaces. The lower limit of V has been studied by K. J. Savage et al. (*28*), who called the limit "plasmon quantum limit". Savage et al. also reported that a remodeling of the plasmons is necessary in the case of a nanogap of dimers below 0.3 nm because surface electrons of a NP can coherently tunnel to another NP during by incident light. In fact, Savage et al. experimentally and theoretically investigated the tunneling effect in sub-nanometer metal gaps and demonstrated that the coherent tunneling effect becomes important at a scale of 0.3 nm for the gaps corresponding to $V = 1.7 \times 10^{-8} (\lambda/n)^3$. The limitation is summarized as (1) the lowest value of the electric field volume and (2) the appearance of new plasmon oscillations of tunneling electrons. These points cannot be derived from the EM mechanism assuming metal surfaces as dielectric surfaces. The interactions among molecular dipoles and such tunneling electrons should be also considered for the optical responses of the molecules near the metal surface within ~0.5 nm, i.e., the Debye radius (*29*).

Interesting Phenomena Appearing at SERS Hotspots beyond the Electromagnetic Mechanism

We introduce here three phenomena occurring when SERS conditions are beyond the limitation of the EM mechanism as follows: (I) ultra-fast SEF, which appears as unique background of SERS (*30*); (II) the plasmon resonance spectra change significantly after losing SERS activity (*31*); and (III) strange Raman lines sometimes appear in the SERS spectra (*32*). These phenomena cannot be explained by the EM mechanism. In the section, we introduce these phenomena with respect to the limitation of the EM mechanism. (I) to (III) can be described as breakdowns of approximations in Fermi's golden rule, which expresses the transition rate (probability of transition per unit time) from one energy eigenstate with a wavefunction $|i\rangle$, of a quantum system to another energy eigenstate with a wavefunction $|f\rangle$, by the perturbation induced by the interaction between light and a molecule as follows (*33*):

$$P \propto \rho\langle i|H'|f\rangle, (3)$$

where P is the transition rate, ρ is the mode density of the field, and $<i|H'|f>$ is the transition element by the perturbation of light H' between the final and initial states. Equation (3) assumes these approximations in the EM mechanism as that (I) the radiative de-excitation rate enhanced by $|M|^2$ becomes slower than the vibrational decay rate, (II) the coupling rate between plasmon resonance and molecular resonance becomes slower than the dephasing rates of both resonance, and (III) the steepened local electric field-gradient at the hotspots is much larger than the size of the molecular electronic structures. The breakdowns in these approximations (I) to (III) are explained as follows.

(I) A product of a fluorescence spectrum and a plasmon resonance spectrum does not correctly reproduce the SEF spectral shape not like SERS, which can be reproduced by a product of a Raman spectrum and a plasmon resonance spectrum as Eq. (3). The failure of reproduction is due to ultra-fast radiative de-excitation by $|M(\lambda_{em})|^2$. When the value of ρ, which is enlarged by $|M(\lambda_{em})|^2$ at the hotspots of silver NP dimers, allows a de-excitation rate larger than the molecular vibrational decay rate in the electronic excited state, the molecule emits light from the vibrational excited state in the electronic excited state. This light emission means that the molecular electronic dynamics deviate from those in a free space. This deviation indicates the breakdown of a well known rule of molecular electronic dynamics; Kasha's rule, in which molecules are assumed to emit light by electronic transition (*34*).

The total radiative decay rates of a molecule in a free space are Γ_{r0} (= $\int\gamma_{r0}(\lambda)d\lambda$), where $\gamma_{r0}(\lambda)d\lambda$ is the decay rates at λ. Vacuum fluctuation yields the decay rates from $|i>$ to $|f>$ due to spontaneous emission (*35*). Let's estimate the ultra-fast emission for the typical SM SERS system composed of the dimer and a R6G molecule located at the dimer junction. The total radiative decay rates enhanced by plasmon resonance are Γ_{Rad} (= $\int\gamma_{r0}(\lambda)|M(\lambda, d_{av})|^2d\lambda$). For a fluorescence molecule, the total internal relaxation rate $\Gamma_{int} \sim 10^{12}$ s^{-1} is far larger than the total radiative decay rate $\Gamma_{r0} \sim 10^{8\text{-}9}$ s^{-1} (*34*). Thus, the fluorescence spectrum of a molecule in a free space is a radiative transition from the bottom of the electronic excited state, which is independent of the excitation laser energy. Under the condition of the conventional SEF, i.e., $\Gamma_{int} >> \Gamma_{Rad}$, the excitation and fluorescence transitions are nearly identical to those for a molecule in a free space. Thus, this type of SEF spectra can be expressed as a product of $\gamma_{nr0}(\lambda)$ and $|M(\lambda, d_{av})|^2$ (*18*, *36*). However, $|M|^2 \sim 10^5$ at hotspots realizes $\Gamma_{int} < \Gamma_{Rad}$. Under this condition, SEF cannot be explained in the same manner as conventional SEF, because it has a component emitting from the vibrational excited state in the electronic excited state, before relaxing to the bottom of the electronic excited state S_1. The emission indicates that the highest energy of the ultra-fast SEF spectra is blue-shifted from the fluorescence of a molecule in a free space and becomes dependent of the excitation laser energy (*36*).

Our experimental works on ultra-fast SEF are here reviewed. We experimentally averaged out the plasmonic modulation in SEF of dye molecules by measuring large Ag NP aggregates which contain a large number of hotspots (*30*). They discovered the excitation laser energy dependence of the ultra-fast

SEF spectra regarding the spectral blue shifts of ~400 meV, the suppression of the anti-Stokes intensity by a factor of several tens, and the super-broadening of the Stokes regions around 1.0 eV as shown in Figure 2. In the framework of the EM effects, the properties are comprehensively explained as the direct emission from the vibrational excited states in S_1 before the electron relaxation to the bottom of S_1 (*5*, *30*).

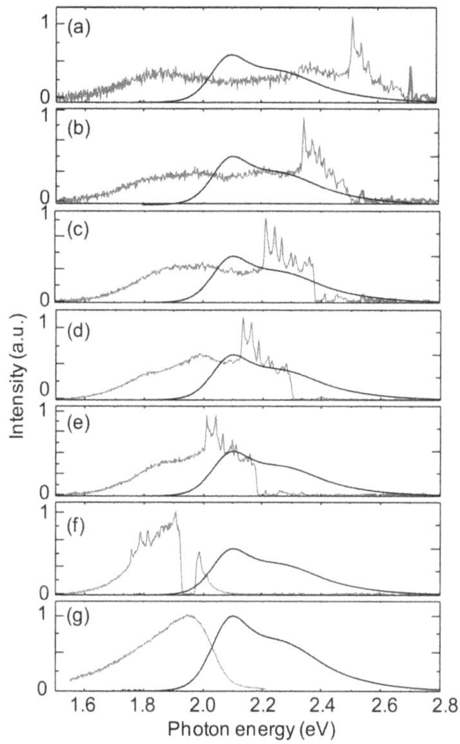

Figure 2. (a–f) Excitation laser energy dependence of SEF spectra of crystal violet (CV) dye from large silver NP aggregates excited with (a) 2.71, (b) 2.54, (c) 2.41, (d) 2.33, (e) 2.21, and (f) 1.96 eV (gray lines) and absorption spectrum of CV molecules in a free space (black lines). Detailed structures around excitation laser lines are SERS bands. (g) Absorption (black line) and fluorescence (gray lines) spectrum of CV molecules in aqueous solution (~10-6 M). Reproduced from Ref. (30). Copyright 2013, American Physical Society.

(II) When the coupling rate between plasmon and molecular excitonic resonances becomes comparable to the dephasing rates of both resonances, both resonance states coherently exchange excitation energy, resulting in the breakdown of the weak coupling approximation. Note that for weak coupling, the approximations of both resonances are independently treated. The ρ and $<i|H'|f>$ interfere with each other, and the two resonances form hybridized resonances. This breakdown is realized by a small electric field volume whose inverse is proportional to the coupling rate, and it can be investigated as a vacuum Rabi splitting by tuning the plasmonic resonance energy (*37*). Note that photon number in a hotspot of a dimer is so small that Rabi splitting by light is too small to generate measureable Ω_R under common SM SERS excitation conditions ~several W/cm^2 (*11*).

The coupling energy $\hbar g$ (g is the coupling rate) is described as an inner product of electric field induced by vacuum fluctuation E_{vac} and the transition dipole moment between the electric states $p_0 = -el_{dip}$ as $\hbar g = |p_0 E_{vac}|$. E_{vac} is derived by the relationship between the zero-point energy and the electric field energy integrating through the mode volume $\int \varepsilon_0 E_{vac}^2 d\omega = (1/2)\hbar\omega$ as $E_{vac} = \sqrt{\frac{\hbar\omega}{2\varepsilon_0 V}}$. Thus, we obtain $g = \sqrt{\frac{p_0^2\omega}{2\varepsilon_0\hbar V}}$ when plasmon and molecular excitonic resonance frequencies are equal to be $\omega\left(=\frac{c}{\lambda}2\pi\right)$. Let's estimate the value of $\hbar g$ for the typical SM SERS system composed of the dimer and a R6G molecule located at the dimer junction. The value of g can be rewritten using oscillator strength (OS) of the local electronic resonance of a molecule $f = 2m\omega l_{dip}^2/(e^2\hbar)$ as (*38*)

$$g = \sqrt{\frac{1}{4\pi\varepsilon_r\varepsilon_0}\frac{\pi e^2 f}{mV}}, \quad (4)$$

where ε_r is the relative permittivity 1.77, f of a R6G molecule is 0.69 (*31*), m is free electron mass ~9.11×10^{-31} kg. Note that vacuum fluctuation just means fluctuation of a space without external fields like light. Thus, we can use the term "vacuum" for g even for any spaces like inside water. Let's try to estimate V for the reported value of splitting of plasmon resonance spectra ~20 meV. One can know that V should be ~1.0×10^{-4} $(\lambda/n)^3$ at $\lambda = 600$ nm and $n = 1.33$. The value of V is enough larger than theoretical minimum $V = 1.7 \times 10^{-8}$ $(\lambda/n)^3$ at the crevasse of the dimer as explained above (*28*), indicating that the reported spectral splitting are reasonably explained by strong coupling between plasmon and excitonic resonances. The effective mode volumes $2.0 \times 10^{-6}(\lambda/n)^3$ are roughly converted into $|M|^2$ of 5.0×10^5 considering low Q of plasmon resonance <10. The estimation implies that SM SERS can be a useful probe to evaluate the strong coupling system.

Our experimental work on the strong coupling is here reviewed. In 2003, We implicitly observed the strong coupling between the plasmons and molecular excitons of R6G located in hotspots of a single dimer (*39*). We implicitly explained the investigation as refractive index effect within the resonance region (*39*, *40*). The explanation is identical to vacuum Rabi splitting as mentioned by Ambjörnsson et al. (*37*). We used a near-SM SERS system as a probe for local fields to quantify observations of strong coupling at the single-molecule level (*31*). During SERS quenching, we found temporal spectral changes in both elastic light scattering and SERS, as shown in Ref. 31. A coupled oscillator model (*41*) composed of a plasmon and a molecular dipole consistently reproduces the changes in the elastic light scattering as shown in Figure 3, indicating that the changes are due to the breaking up of the strong coupling by molecular dissipation. These investigations indicate that SM SERS systems are useful for probing dynamics of strong coupling (*5*, *31*).

(III) When the steepness of local electric field-gradient at the hotspots becomes comparable to the steepness of a dipole field of the molecular electronic transition, the excitation of multi-poles such as a quadrupole, which is negligible in most cases of excitation by light in a free space, is significant. Owing to this unique excitation property, the perturbation term induced by H' cannot be approximated by a linear function, resulting in changes of the integrated value $\langle i|H'|f\rangle$ that determines the selection rule for the molecular electronic excitation. This failure of the approximation of the linear function indicates the breakdown of the selection rule. This breakdown has been investigated by an observation of Raman forbidden transition modes (*5*).

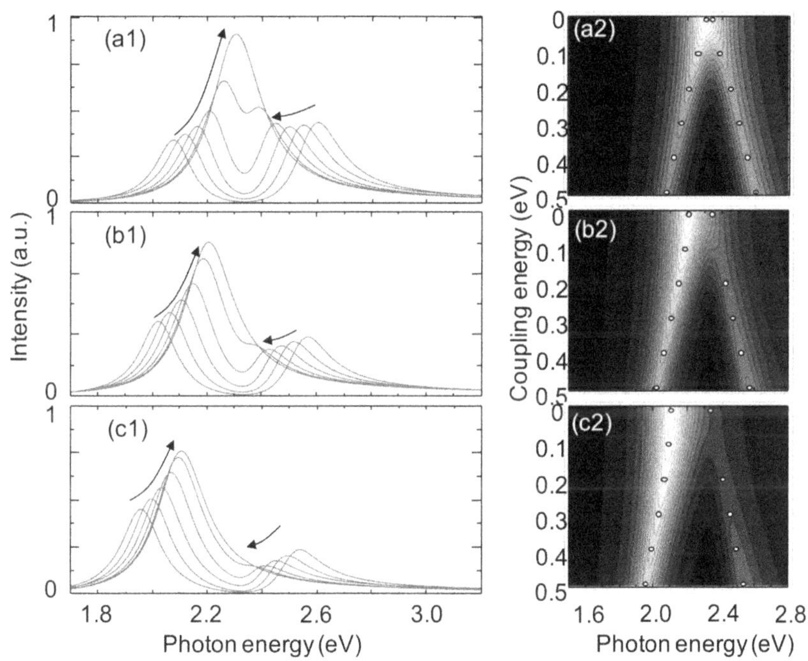

Figure 3. Calculated plasmon resonance spectra using ħg from 500 to 0 meV with interval of 100 meV. The values of original plasmonic resonance energy of (a1)-(c1) are 2.3, 2.2, 2.1 eV, respectively. The value of original line-width of (a1)-(c1) is 220 meV. (a2)-(c2) Contour images of the changes in plasmon resonance spectra. Peak positions of lower and higher energy lines are indicated by circles. Reproduced from Ref. (31). Copyright 2014, American Physical Society.

Theoretical aspects of the breakdown of the selection rule by the local electric field-gradient were developed by several groups using Buckingham's work (*42*, *43*). This breakdown was expressed using the transition dipole moment of the excitons without approximating H' in Eq. (1) as a linear function. Here, we use explanation by E. J. Ayars et al using Buckingham's theory (*32*). The transitions in vibration levels due to coupling with an electric field are described by the perturbation Hamiltonian in Eq. (3) $H' = -\mu E$, where $\boldsymbol{\mu}$ is the dipole moment, and E is the electric field of light. The dipole moment induced in a molecule placed in E_{loc} is given as

$$\mu = \mu^{p} + \alpha E_{\mathrm{loc}} + \cdots, (5)$$

where μ^{p} is the permanent dipole moment and α is the electric dipole-dipole polarizability (*42*). Ayars et al. described that the lowest order of the field-gradient effect contributes to the second term in Eq. (5). Vibrational spectroscopy obtains signals from the fluctuating dipole moment related to the coordinate of the nucleus Q_{n}. Thus, using partial differentiation, Eq. (5) is expanded by Q_{n} and the terms related to Q_{n} are collected as

$$\mu = \left(\left(\frac{d\mu^{\mathrm{p}}}{dQ_{\mathrm{n}}} \right)_{0} + \left(\frac{d\alpha}{dQ_{\mathrm{n}}} \right)_{0} E_{\mathrm{loc}} + \alpha \left(\frac{dE_{\mathrm{loc}}}{dQ_{\mathrm{n}}} \right)_{0} + \cdots \right) Q_{\mathrm{n}} \cdot (6)$$

The terms without Q_{n} are discarded because we cannot observe them by vibrational spectroscopy. The first and second terms in Eq. (6) are infrared absorption (IR) and Raman, respectively. The third term is called the field gradient (FG) Raman term and may be important at the hotspot, because $\left(\frac{dE_{\mathrm{loc}}}{dQ_{\mathrm{n}}} \right)_{0}$ in Eq. (6) may not be negligible in SERS hotspots where the steepness of local electric field-gradient is quite large. A FG term of a propagating electric field in a free space $E^{\mathrm{in}}\exp i((2\pi/\lambda)z+\omega t)$ is obtained as $i2\pi/\lambda$. Thus, in a free space, the ratio of the FG Raman term to the Raman term becomes the order of $2\pi l_{\mathrm{dip}}/\lambda$, where l_{dip} ~0.1 nm for a R6G molecule (*24*) and a wavelength of ~500 nm, respectively. Thus, this ratio is ~10^{-3}, and the FG effect is insignificant. The FDTD calculations of the EM fields at a SERS hotspot suggest that the intensity of the local electric fields changes by ~50% within 0.5 nm (*11*). Thus, this ratio is ~10^{-1} for a R6G molecule, indicating that the breakdown of the selection rule in SERS may be still negligible because much weaker than normal Raman lines. According to Eq. (6), the FG Raman signal should be large when the polarizability is large resulting in a similar selection rule to IR spectroscopy.

Recently, Takase et al. demonstrated that the FG effect can induce forbidden electronic transitions (not vibrational transitions) in SERS signals from carbon nanotubes and observed a forbidden resonance Raman transition using carbon nanotubes (*44*). Sun et al demonstrated that forbidden vibrational transitions at nanogaps by high vacuum tip-enhanced Raman spectroscopy (HV-TERS) (*45*). These works show that the nanogap enhances not only the EM-field intensity but also increases the optical excitation channels, including the excitation of forbidden modes beyond the selection rules. The demonstration of the importance of the FG effect will pave the way for unique applications of SERS.

Summary and Outlook

In this chapter, we first summarized the elucidation of the EM mechanism using SERS phenomena. Note that the analysis of the relations among the SERS spectra, plasmon resonance spectra, and shapes of silver NP dimers is indispensable for the quantitative evaluation of the SERS EM enhancement mechanism (*11*). We second discussed the limitations of the EM mechanism in SERS. Note that nonlinear optical response of molecules and unscreened surface electrons of metal determine the limitation of the electromagnetic mechanism (*24*). We third discussed the necessity for re-examining the EM mechanism in SERS with respect to three breakdowns in the approximations of Fermi's golden rule: the breakdown of the molecular electronic dynamics by the ultra-fast plasmonic de-excitation (*30*); the breakdown of the weak coupling between the plasmon and molecular electron by the strong EM enhancement (*31*); and the breakdown of the selection rule of SERS by the FG effect (*32*). We expect that these breakdowns will allow the creation of novel plasmonic spectroscopic fields including unique chemical reactions on plasmonic metal surfaces (*46, 47*). We expect that contribution of chemical enhancement in plasmon enhanced spectroscopy will be also important in this new research field (*48*). We believe that such breakdowns can contribute to various SERS applications including development of SERS substrates and ultra-sensitive detection of bio-related molecules (*49–51*).

Acknowledgments

This work was supported by the Japan Society for the Promotion of Science (JSPS) KAKENHI Grant-in-Aid for Scientific Research (C) Number 20510111, (B) Number 26286066.

References

1. Wang, D. S.; Kerker, M. *Phys. Rev. B* **1981**, *24*, 1777–1790.
2. Inoue, M.; Ohtaka, K. *J. Phys. Soc. Jpn.* **1983**, *52*, 3853–3864.
3. Aravind, P. K.; Nitzan, A.; Metiu, H. *Surf. Sci.* **1981**, *110*, 189–204.
4. Ueno, K.; Misawa, H. *J. Photochem. Photobiol. C* **2013**, *15*, 31–52.
5. Yamamoto, Y. S.; Ozaki, Y.; Itoh, T. *J. Photochem. Photobiol. C* **2014**, *21*, 81–104.
6. Moskovits, M. *J. Chem. Phys.* **1978**, *69*, 4159–4161.
7. Bohren, C. F.; Huffman, D. R. *Absorption and Scattering of Light by Small Particles*; Wiley: New York, 1983.
8. Nie, S.; Emory, S. R. *Science* **1997**, *275*, 1102–1106.
9. Kneipp, K.; Wang, Y.; Kneipp, H.; Perelman, L. T.; Itzkan, I.; Dasari, R. R.; Feld, M. S. *Phys. Rev. Lett.* **1997**, *78*, 1667–1670.
10. Xu, H.; Aizpurua, J.; Kall, M.; Apell, P. *Phys. Rev. E* **2000**, *62*, 4318–4324.
11. Yoshida, K.; Itoh, T.; Tamaru, H.; Biju, V.; Ishikawa, M.; Ozaki, Y. *Phys. Rev. B* **2010**, *81* (115406).
12. Itoh, T.; Asahi, T.; Masuhara, H. *Appl. Phys. Lett.* **2001**, *79*, 1667–1669.

13. Itoh, T.; Hashimoto, K.; Ozaki, Y. *Appl. Phys. Lett.* **2003**, *83* (2274).
14. Itoh, T.; Biju, V.; Ishikawa, M.; Kikkawa, Y.; Hashimoto, K.; Ikehata, A.; Ozaki, Y. *J. Chem. Phys.* **2006**, *124* (134708).
15. Sun, G.; Khurgin, J. B. *Phys. Rev. A* **2012**, *85* (063410).
16. Itoh, T.; Hashimoto, K.; Ikehata, A.; Ozaki, Y. *Chem. Phys. Lett.* **2004**, *389*, 225–229.
17. Itoh, T.; Yoshida, K.; Biju, V.; Kikkawa, Y.; Ishikawa, M.; Ozaki, Y. *Phys. Rev. B* **2007**, *76* (085405).
18. Yoshida, K.; Itoh, T.; Biju, V.; Ishikawa, M.; Ozaki, Y. *Phys. Rev. B* **2009**, *79* (085419).
19. Yoshida, K.; Itoh, T.; Biju, V.; Ishikawa, M.; Ozaki, Y. *Appl. Phys. Lett.* **2009**, *95* (263104).
20. Itoh, T.; Yoshikawa, H.; Yoshida, K.; Biju, V.; Ishikawa, M. *J. Chem. Phys.* **2009**, *130* (214706).
21. Itoh, T.; Yoshikawa, H.; Yoshida, K.; Biju, V.; Ishikawa, M. *J. Chem. Phys.* **2010**, *133* (124704).
22. Itoh, T.; Iga, M.; Tamaru, H.; Yoshida, K.; Biju, V.; Ishikawa, M. *J. Chem. Phys.* **2012**, *136* (024703).
23. Kneipp, K.; Wang, Y.; Kneipp, H; Itzkan, I; Dasari, R. R.; Feld, Michael S. *Phys. Rev. Lett.* **1996**, *76*, 2444–2447.
24. Johansson, P.; Xu, H.; Käll, M. *Phys. Rev. B* **2005**, *72* (035427).
25. Xu, H.; Wang, X-H.; Persson, M. P.; Xu, H. Q.; Käll, M.; Johansson, P. *Phys. Rev. Lett.* **2004**, *93* (243002).
26. Scully M. O.; M. Suhail Zubairy, M.; *Quantum Optics,* Cambridge University Press, Cambridge, 1997.
27. Larkin, I. A.; Stockman, M. I.; Achermann, M.; Klimov, V. I. *Phys. Rev. B* **2004**, *69* (121403).
28. Savage, K. J.; Hawkeye, M. M.; Esteban, R.; Borisov, A. G.; Aizpurua, J.; Baumberg, J. J. *Nature* **2012**, *491*, 574–577.
29. van Hulst, N. F. *Nat. Nanotechnol.* **2012**, *7*, 775–777.
30. Itoh, T.; Yamamoto, Y. S.; Tamaru, H.; Biju, V.; Murase, N.; Ozaki, Y. *Phys. Rev. B* **2013**, *87* (235408).
31. Itoh, T.; Yamamoto, Y. S.; Tamaru, H.; Biju, V.; Wakida, S.; Ozaki, Y. *Phys. Rev. B* **2014**, *89* (195436).
32. Ayars, E. J.; Hallen, H. D.; Jahncke, C. L. *Phys. Rev. Lett.* **2000**, *85*, 4180–4183.
33. Dirac, P. A. M. *Proc. R. Soc. A* **1927**, *114*, 243–265.
34. Kasha, M. *Disc. Faraday Soc.* **1950**, *9*, 14–19.
35. Sakurai, J. J.; *Advanced Quantum Mechanics*; Addison-Wesley: Reading, MA, 1967.
36. Le Ru, E. C.; Etchegoin, P. G.; Grand, J.; Fe´lidj, N.; Aubard, J.; Le´vi, G. *J. Phys. Chem. C* **2007**, *111*, 16076–16079.
37. Ambjörnsson, T.; Mukhopadhyay, G.; Apell, S. P.; Käll, M. *Phys. Rev. B* **2006**, *73* (085412).
38. Andreani, L. C.; Panzarini, G.; Gerard, J. M. *Phys. Rev. B* **1999**, *60*, 13276–13279.

39. Itoh, T.; Hashimoto, K.; Ikehata, A.; Ozaki, Y. *App. Phys. Lett.* **2003**, *83*, 5557–5559.
40. Itoh, T.; Kikkawa, Y.; Biju, V.; Ishikawa, M.; Ikehata, A.; Ozaki, Y. *J. Phys. Chem. B* **2006**, *110*, 21536–21544.
41. Wu, X.; Gray, S. K.; Pelton, M. *Opt. Express* **2010**, *18*, 23633–23645.
42. Buckingham, A. D. *Adv. Chem. Phys.* **1967**, *12*, 107–142.
43. Sass, J. K.; Neff, H.; Moskovits, M.; Holloway, S. *J. Phys. Chem.* **1981**, *85*, 621–623.
44. Takase, M.; Ajiki, H.; Mizumoto, Y.; Komeda, K.; Nara, M.; Nabika, H.; Yasuda, S.; Ishihara, H.; Murakoshi, K. *Nat. Photon.* **2013**, *7*, 550–554.
45. Sun, M.; Zhang, Z.; Chen, L.; Li, Q.; Sheng, S.; Xu, H.; Song, P. *Adv. Mater. Interface* **2014**, *1* (1300125).
46. Yamamoto, Y. S.; Hirano, K.; Ishido, T.; Yasui, T.; Murase, N.; Baba, Y.; Itoh, T. *Phys. Chem. Chem. Phys.* **2013**, *15*, 10316–10320.
47. Itoh, T.; Yamamoto, Y. S.; Biju, V.; Tamaru, H.; Wakida, S. *AIP Adv.* **2015**, *5* (127113).
48. Yamamoto, Y. S.; Itoh, T. *J. Raman Spectrosc.* **2016**, *47*, 78–88.
49. Yamamoto, Y. S.; Hasegawa, K.; Hasegawa, Y.; Takahashi, N.; Kitahama, Y.; Fukuoka, S.; Murase, N.; Baba, Y.; Ozaki, Y.; Itoh, T. *Phys. Chem. Chem. Phys.* **2013**, *15*, 14611–14615.
50. Yamamoto, Y. S.; Fujime, Y.; Takahashi, N.; Nakanishi, S.; Itoh, T. *RSC Adv.* **2016**, *6*, 31454–31461.
51. Yamamoto, Y. S.; Ishikawa, M.; Ozaki, Y.; Itoh, T. *Front. Phys.* **2014**, *9*, 31–46.

Chapter 3

Patterning Surface-Enhanced Raman Scattering Substrates Obtained Using the Porous Character of Colloidal Microsphere Monolayer Templates

Weidong Ruan,[1,2] Tieli Zhou,[3] Xu Wang,[2] Young Mee Jung,[*,1] and Bing Zhao[2]

[1]Department of Chemistry, Institute for Molecular Science and Fusion Technology, Kangwon National University, Chunchon 24341, Korea
[2]State Key Laboratory of Supramolecular Structure and Materials, Jilin University, Changchun 130012, P. R. China
[3]College of Food Science and Engineering, Changchun University, Changchun 130012, P. R. China
[*]Tel: +82 33 250 8495. Fax: +82 33 259 5667.
E-mail: ymjung@kangwon.ac.kr

Surface-enhanced Raman scattering (SERS) is one of the most noted applicable properties of plasmonic materials. Fabricated SERS substrates, normally including Ag and Au nanostructures, are equivalent to plasmonic materials in terms of their enhanced spectroscopic properties. The enhancement of the vibrational modes of molecules dwelling in/on plasmonic materials is the main issue of concern for SERS. In this chapter, patterning techniques for SERS substrates, obtained by using the porous character of polystyrene (PS) colloidal microsphere monolayer templates, are introduced. Several newly developed strategies that mainly concentrate on the filtration effect of the PS templates are considered. The uniform micro/nano pores of the PS templates serve as pathways for the coating of metal nanoparticles (NPs). Through a combination of the layer-by-layer self-assembly technique for polyelectrolytes and the mask effect of PS templates, micro/nano-patterned films, including nanowell, nanoring, and nanodot arrays, are formed. Methods to produce compound structures of NP-decorated

PS microspheres with two components are illustrated. These nanostructures are composed of Ag, Au, or both. Sputtering and co-sputtering techniques are also introduced for multi-material deposition based on the PS templates. The SERS properties of a few typical nanostructures, composed of the same building blocks of Ag and Au NPs but in different aggregate patterns, are reviewed. Additionally, the mode by which the aggregate status of plasmonic materials finally affects the enhancement ability for SERS is shown as a proof-of-concept.

Introduction

Currently, there is great interest in the synthesis and applications of complex nanostructures with periodic patterns made by template techniques such as the photo-lithographic approach (*1*), biomolecular self-assembly (SA) (*2*), breath figures approach (*3*), capillary force lithography (*4*), and so on. The preparation of nanostructured surfaces with defined compositions and morphologies is of immense scientific and technological interest.

Among the template techniques, the use of polystyrene (PS) microspheres has attracted extensive interest in recent years due to their perfect and spontaneously formed periodic structures (*5*). In combination with other techniques, the obtained microsphere arrays can be used as versatile templates to fabricate complex nanostructures (*6–9*). The employed methods include thermoevaporation, electrochemistry, chemical deposition, and ion beam etching, among others. A rich array of functionalities, such as optical activity, mechanical strength, magnetism, and conductivity, is provided by the processed microspheres, which are used as advanced functional materials. The so-called colloidal lithography has numerous applications in material science.

The plasmonic behavior is one of the most renowned applications of these nanomaterials. Enhanced spectroscopy, especially surface-enhanced Raman scattering (SERS), is the most famous application. SERS is a special spectroscopic phenomenon in which weak signals can be enhanced by several orders of magnitude. It has evolved into a powerful and reliable analytical tool for ultrasensitive detection of analytes. For the successful application of SERS, the fabrication of substrates is of utmost importance. In a wide variety of SERS-active architectures, ordered substrates are always desirable because they provide the advantage of stable and reproducible signal acquisition. This chapter describes the fabrication of patterned SERS substrates using PS microsphere templates. The obtained nanostructures include micropore, nanodot, and nanoring array films composed of Ag, Au, or both. The filtration effect of PS templates is the core strategy used in the fabrication procedures.

PS microspheres are mainly used here as the template material due to their advantage of easy elimination by either the physical method of plasma etching or the chemical method of dissolution in organic solvents. Monolayered PS templates were focused and combined with the layer-by-layer (LbL) SA method. It is well known that LbL SA is a versatile strategy for surface modification and/or coating in wet chemistry. The versatility of this strategy provides great potential to create a variety of micro/nano structures. Here, Au and/or Ag nanoparticles (NPs) were mostly employed as building blocks, due to their distinguished plasmonic properties.

The literature reports many methods for processing PS templates, including ion beam etching (*10*), prefabricated guiding templates (*11*), deformable soft lithography (*12*), etc. In this chapter, we focus on a patterning technique based on raw PS templates (as-prepared PS templates). It should note that the mentioned methods can also be combined with other processing techniques for PS templates, resulting in more complex micro/nano structures and potential applications in nanoscience and nanotechnology. The introduced ideas provide insight into the fabrication of new patterned substrates. The following sections will provide information about various pattering techniques for SERS substrates. A detailed discussion also covers the application in SERS detection.

Patterns of Nanowell Arrays Composed of NP Films

A vertical lifting method, the Langmuir–Blodgett (LB) technique (*13*, *14*), is employed to prepare the PS monolayer templates. An illustration of the employed LB process is shown in Figure 1.

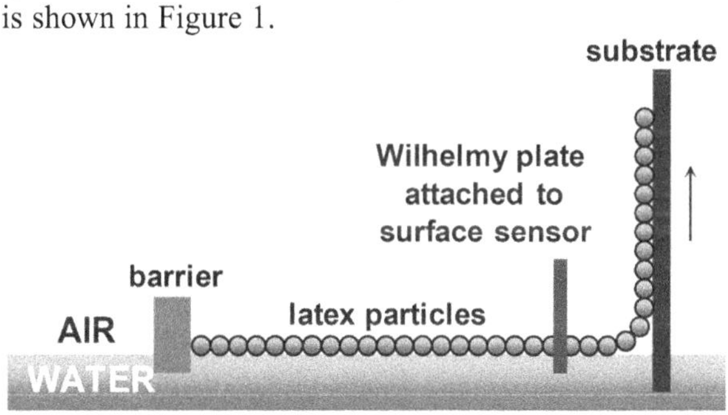

Figure 1. Schematic illustration of the LB deposition technique used to assemble PS microspheres on a planar substrate. (Reproduced with permission from reference (13). Copyright 2007 Elsevier).

The basic micropatterns of nanowell arrays composed of NPs films are described here. Based on the nanosphere lithography (NSL) (*5*) and LbL SA techniques, periodic micro/nano structures of the contact areas between PS templates and glass or silica substrates are faithfully replicated on the substrate surface. Figure 2a shows a typical SEM image of a PS monolayer template with a large-area ordered array of PS microspheres. The polyelectrolyte of poly(diallyldimethylammonium chloride) (PDDA) is employed as an adhesive agent in the LbL process. Figure 2b shows the SEM image of as-prepared PS films coated with Au NPs. The ordered nanowells exactly replicate the period of the PS template. Due to the versatile applications of the LbL SA technique, other types of NPs can also be deposited as equivalent substitutes. This is the general method for the construction of patterned surfaces (*15*).

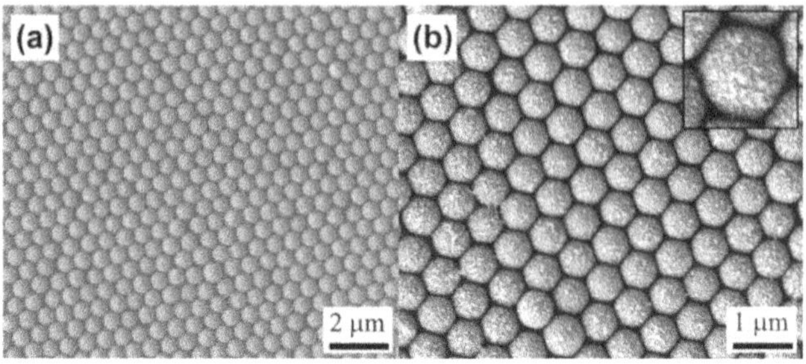

Figure 2. SEM image of (a) a two-dimensional (2D) PS microsphere array in a hexagonal close-packed pattern on a glass substrate, and (b) SEM images of a monolayer of PS microspheres coated with a layer of Au NPs. (Reproduced with permission from reference (15). Copyright 2010 Royal Society of Chemistry).

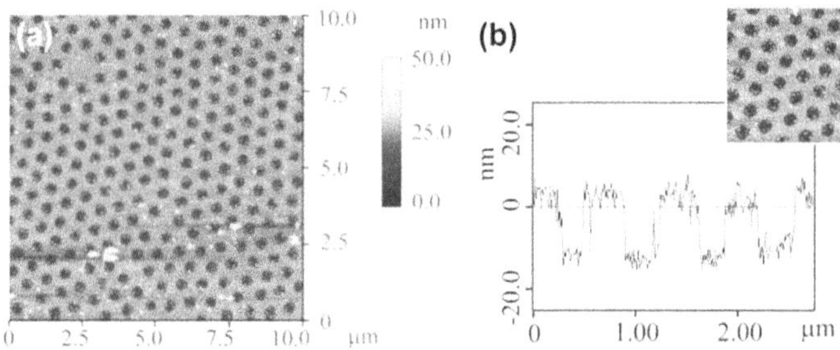

Figure 3. (a) AFM image of a Au nanowell array consisting of Au NPs (ave. 17 nm) on the glass slides. (b) A cross-sectional contour of the Au nanowell array. The diameter and depth of the wells are approximately 325 and 17 nm, respectively. (Reproduced with permission from reference (15). Copyright 2010 Royal Society of Chemistry).

Patterns on glass substrates are obtained by removing the PS templates. The atomic force microscopy (AFM) image in Figure 3a clearly shows the obtained periodic Au nanowell arrays. A line scan, such as the one shown in Figure 3b, can be used to determine the geometry of the nanowells. In this case, the nanowells have an average diameter of approximately 325 nm and an average depth of 17 nm (*15*). By combining the NSL and LbL SA techniques, Au nanowell arrays can easily be obtained, as demonstrated above. This is a unique technique for surface functionalization with periodic patterning. This effective method provides an opportunity to produce surface modifications with applications in SERS analysis.

Particle-Lithography-Based Patterning of Polyelectrolyte Films for Deposition of NPs

The NSL technique combined with the LbL SA technique also provides an opportunity for site-specific nanoscale deposition of NPs. The following discussion demonstrates a simple yet effective method to fabricate different types of surface micropatterns with polyelectrolytes (PEs) or composites of PEs and metallic NPs via a particle lithography method.

PEs are a type of polymer with many applications in sol-gel chemistry, especially to impart surface charges for the subsequent assembly of NPs (*16–19*). The basic strategy is to create patterned PE templates (PPTs). Two synthetic routes were designed to achieve the desired nanostructures. In the first route, films of one molecular layer of PPT were fabricated on surfaces and were subsequently used to guide the deposition of Au NPs. In the second route, a complicated pattern of Au NP films with inlaid circular PE dot microarrays was fabricated, followed by the deposition of the Ag NPs via site-specific adsorption.

Figure 4 illustrates the fabrication procedure for the PPT films and the corresponding NP assemblies. Figure 4a shows a schematic representation of route 1, which involves a four-step synthesis process. The crucial step is step 3, which involves the site-selective adsorption of Au NPs on the PPT films and the formation of the intended patterned nanostructures. The quaternary ammonium cations present in the main chain of PDDA provide abundant positive charges to the PPT films, which can adsorb the negatively charged Au NPs effectively. However, the bare glass surfaces are electrically neutral and cannot adsorb the Au NPs. Thus, the circular vacant microarrays are left on the surfaces. Fabrication through route 2 (Figure 4b) involves five steps. The unique dual component patterns of Au and Ag NP films were obtained after a five-step synthesis. Before the final step to adsorb Ag NPs, the areas for the Ag NP deposition were protected by PS microspheres in step 3 and exposed again in step 4. This design helped in obtaining the desirable Au–Ag compound micro/nanostructures (*20*). Particle lithography was used for fabricating PPT films via both routes 1 and 2. In route 1, selective exposure sites were created in a single step in PE films for the adsorption of NPs. However, in route 2, step-by-step (two-step) selective exposures of PE films were performed for the deposition of different NPs. Thus, the employed method exhibited versatility for creating PPT films and assembling the NPs (*20*).

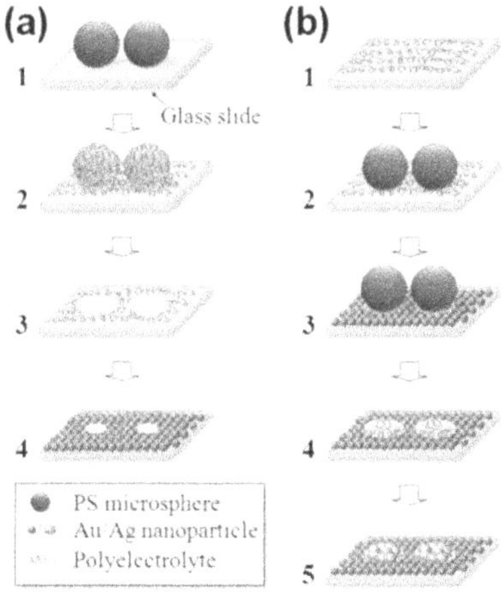

Figure 4. Schematic illustration of the fabrication procedure for PPT films and the corresponding Au/Ag NP assemblies. (a) Schematic representation of route 1. (b) Schematic representation of route 2. (Reproduced with permission from reference (20). Copyright 2014 Elsevier).

Figure 5 shows the AFM images of the samples obtained from route 1. The top view (Figure 5a) and the corresponding line scan (Figure 5b) show that the PPT films obtained in step 3 were uniform and extremely smooth. No obvious patterns were observed because of the insufficient resolution of the AFM on monolayered PPT films. Figures 5c and d reveal that the adsorption of Au NPs on the PPT films led to the formation of a distinct pattern. The vacant sites have an average top width of 130 nm and an average depth of 17 nm, corresponding to the thickness of one layer of Au NPs (*20*).

The SEM images of the samples obtained in steps 4 and 5 of route 2 are shown in Figure 6. The patterned NP films faithfully replicated the periodic arrays of PS microspheres. The nanowells possessed a monolayer of PE films with the potential to adsorb extra NPs. Figure 6b shows the SEM image after the deposition of extra NPs of Ag. The vacant circular nanowell arrays are covered exactly by the Ag NPs (45 nm) exactly. Binary component nanofilms were obtained, composed of continuous films of Au NPs and dot microarrays of Ag NPs. Particle lithography showed great flexibility in fabricating the patterned PPT films and the corresponding NP assemblies (*20*).

In a further exploration, the edges of the nanowells were developed to selectively adsorb Ag NPs. The preparation of necklace-like array films of Ag NPs is a modification of route 2 in Figure 4. Five steps were employed (*21*), as follows: (i) deposit a template monolayer of PS microspheres (600 nm) on the glass slides; (ii) deposit a layer of PDDA on the PS template; (iii) deposit Au NPs on the PDDA film; (iv) eliminate the PS monolayer template; and (v) deposit Ag NPs on the Au NP precursor film.

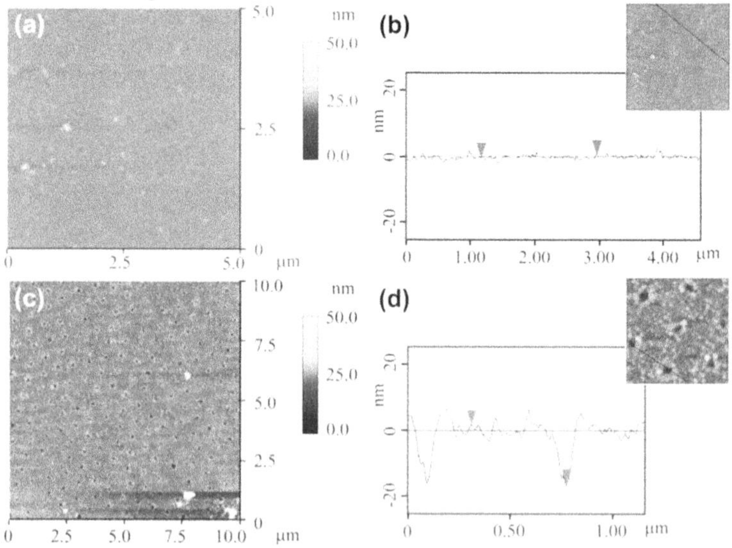

Figure 5. AFM images of the samples obtained from route 1. (a) The top view of the obtained PPT films; (b) a corresponding line scan of the obtained PPT films; (c) the top view of the Au NP films with inlaid nanowell microarrays; and (d) a corresponding line scan of the Au NP films with inlaid nanowell microarrays. (Reproduced with permission from reference (20). Copyright 2014 Elsevier).

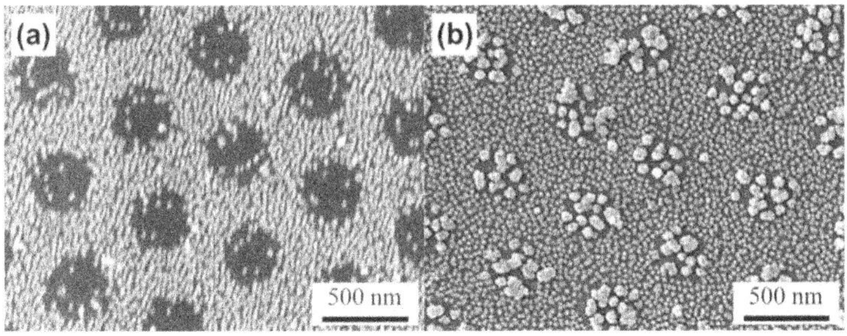

Figure 6. SEM images of the morphologies of samples obtained from steps 4 and 5 in route 2. (a) SEM image of the nanowell arrays of Au NP films. (b) SEM image of the binary-component nanofilms, consisting of continuous Au NP films with inlaid Ag NP dot microarrays. (Reproduced with permission from reference (20). Copyright 2014 Elsevier).

The reason for the site-selective adsorption of Ag NPs on the edges of the nanowell array films of Au NPs is the abundance of PDDA dwelling at the edges of the periodic nanowells. After the elimination of the PS templates, a great deal of polyelectrolyte is exposed on the edges. The exposed polyelectrolyte is very active and can be used to form new nucleation sites for a secondary electrostatic assembly. This is the basis for the deposition of Ag NPs on the edges. A so-called surface-guided SA technique was created (*21*), which was the first report of surface-guided SA on the edges of covered polyelectrolyte films.

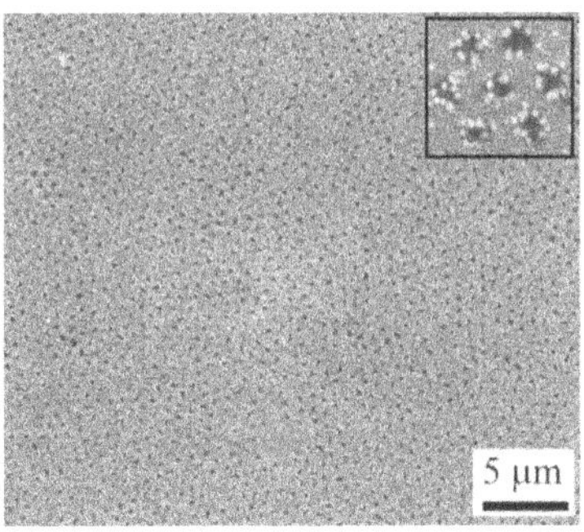

Figure 7. SEM image of a Ag "necklace" array on a pre-patterned substrate. (Reproduced with permission from reference (21). Copyright 2008 American Chemical Society).

Figure 7 shows the SEM image of the samples after the selective deposition of Au NPs on the nanowell films. A "necklace" structure was formed along the edges of the nanowells. These nanostructures could be used as substrates for plasmonics or SERS.

Sputtering and Co-Sputtering Technique for Multi-Material Deposition Based on PS Templates

The sputtering technique is one of the most important techniques for the preparation of SERS-active substrates. Both LbL sputtering and co-sputtering of different materials were employed for SERS substrate preparation based on the PS template. LbL sputtering and co-sputtering techniques for multiple materials in different layers and the same layer were employed for evaluation of SERS enhancement. The compositions of the materials, and their structure and thickness, all of which affect the SERS activity, were dependent on the sputter time and power. Previous studies of the SERS activity have been performed with varying material compositions, structures, and thicknesses.

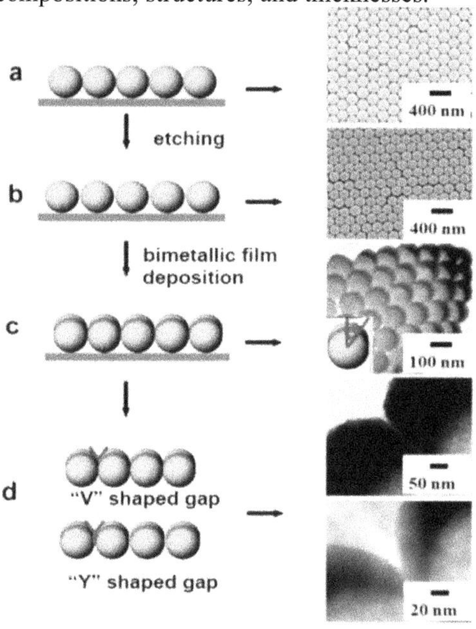

Figure 8. Schematic, SEM, and TEM images representing the fabrication of Au/Ag bilayer nanocap arrays. (a) Close-packed PS array. (b) Separated PS array. (c) Au/Ag nanocap array on PS template. The inset is the schematic of the film thickness for the bilayer on the PS sphere surface. (d) "Y"-shaped nanogap and "V"-shaped junction. (Reproduced with permission from reference (23). Copyright 2015 Royal Society of Chemistry).

Au and Ag materials are good candidates for SERS studies. When the original PS array is used as a substrate for Ag deposition, SERS enhancements show strongly size-dependent behavior. When O_2-plasma is introduced to etched 2D PS templates, nanogaps are formed with the deposition of Ag. The size of the nanogaps can be controlled from 5 to 20 nm, which is the perfect scale for SERS. The significant SERS effects are due to the coupling between Ag nanocaps accompanied with the rough surface of the array, which enhances the local electric-field intensity (*22*). Another method for the fabrication of SERS substrates is sputtering of the Au and Ag layers to form an Au/Ag bilayer (Figure

8). With the deposition of the Au and Ag layers, "Y"-shaped nanogaps and "V"-shaped junctions can be tailored between neighboring nanogaps by changing the bilayer thickness and PS diameter. The proposed Au/Ag bilayer exhibited significantly higher SERS enhancement than either the Ag or Au monolayers on the PS template. The electromagnetic coupling between the Au and Ag bilayers and the nanogaps generates extremely intense local electromagnetic fields, and the composite provides more possibilities to accommodate probing molecules (*23*).

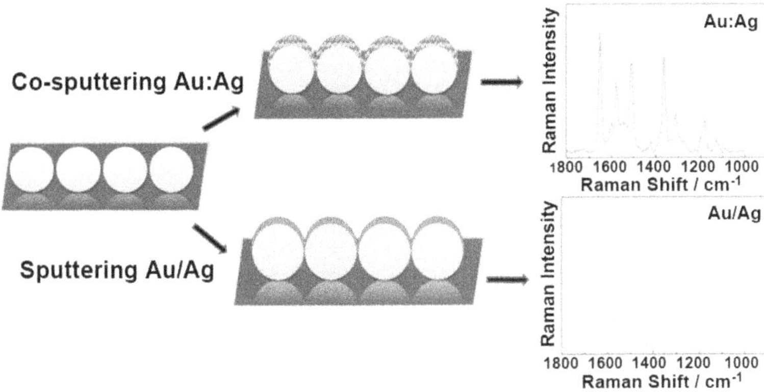

Figure 9. Schematic of co-sputtered Au:Ag bimetal and sputtered Au/Ag bilayer on a PS template array, and SERS spectra of R6G adsorbed on the co-sputtered Au:Ag bimetal and sputtered Au/Ag bilayer. (Reproduced with permission from reference (24). Copyright 2016 Elsevier).

As shown in Figure 9, Au:Ag bimetal nanocap arrays with a waxberry-like structure were previously reported by co-sputtering Au and Ag onto 2D PS colloidal sphere templates, based on a magnetron sputtering system. The Au:Ag bimetal array with a waxberry-like structure formed the protruding network of Ag and Au NPs, which contributed to higher SERS enhancement. Compared to the bilayer Au/Ag, the co-sputtered Au:Ag bimetal array exhibited 4-fold greater enhancement of the SERS signal intensity of Rhodamine 6G (R6G), which was ascribed to the plasmon coupling between the NPs of Au and Ag on the PS array (*24*). Additionally, the Au:Ag bimetal provides an abundance of "hot spots" by the formation of metal layers from two different metals (Au and Ag). The enhancement factor (*EF*) of the Au:Ag bimetal nanocap array substrate on which R6G was adsorbed was evaluated as 6.72×10^{10}.

The nanocap-shaped SiO_2-isolated Ag array was prepared based on the PS template and co-sputtering technique for the deposition of Ag and SiO_2, in which hot spots could be engineered in a facile manner on a three-dimensional (3D), curved surface. The SERS activities of the SiO_2-isolated Ag array vary non-monotonically and depend strongly on the film thickness and surface roughness. In this system, the greatest advantage of the addition of SiO_2 is not only that it avoids photobleaching and background fluorescence but also that it decreases the rate of oxidation of Ag and increases the stability of Ag particles. More importantly, the addition of SiO_2 improves the enhancement of the electrical field at the nanocap arrays. Figure 10 shows the fabrication process for a 2D array of PS by SA and co-sputtering techniques. The SEM and TEM images show the colloidal sphere array covered with a SiO_2-isolated Ag layer (Figure 10) (*25*). Based on previous research of SiO_2-isolated Ag arrays, "pillar-cap" shaped Ag/SiO_2 multilayers were also developed on the PS template by the co-sputtering of Ag and SiO_2 materials. In the proposed system, 2–5 nm silver NPs were squeezed into the SiO_2 sublayer through its pinholes. The size of the Ag NPs is strongly dependent on the annealing temperature. Based on the co-sputtering and annealing techniques, "hot spots" were abundantly formed for SERS. This method provides multi-diffusion interfaces between Ag and SiO_2 hybrid structures.

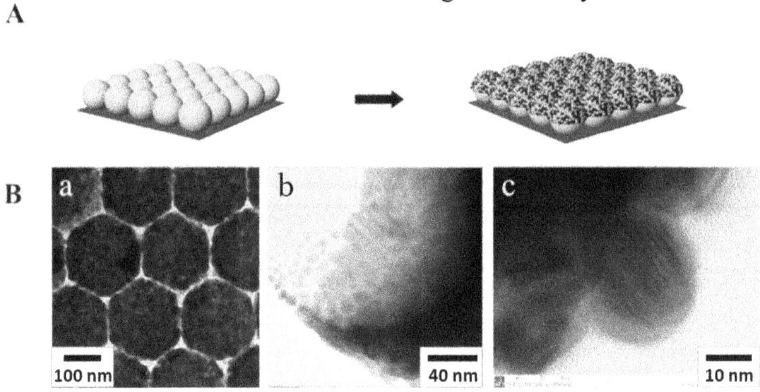

Figure 10. (A) Fabrication process of 2D arrays of PS by SA on a Si wafer. Ag and SiO_2 are co-sputtered along the perpendicular direction onto the close-packed colloid sphere arrays. (B) TEM and HRTEM images of the SiO_2−Ag nanocap arrays on a 200 nm PS template for the SiO_2−Ag film with 40 nm thickness. (a) TEM images of the SiO_2−Ag nanocap arrays on a 200 nm PS template, (b) and (c) TEM image of a single SiO_2−Ag nanocap. (Reproduced with permission from reference (25). Copyright 2014 American Chemical Society).

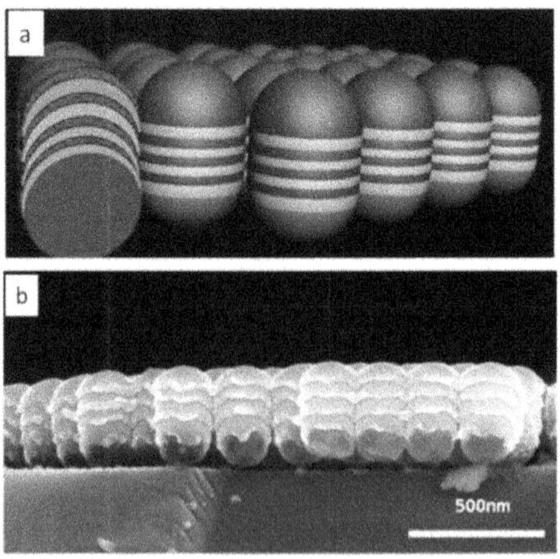

Figure 11. (a) Schematic diagram of the structure of the pillar-cap Ag/SiO_2 multilayer array. (b)Section-view SEM images for the as-fabricated Ag/SiO_2 multilayer array. (Reproduced with permission from reference (26). Copyright 2016 Elsevier).

Figure 11 presents a schematic diagram of the structure of the pillar-cap array, and the SEM image of the profile view of the as-fabricated multilayer array (*26*). The proposed technique opens a new way to prepare multi-materials co-sputtered with arbitrary composites of metals (Au, Ag, Pd, and Pt) and semiconductors (TiO_2, ZnO, ZnS and CuO).

Surface Decorations of PS Microspheres Based on Size-Dependent Filtration of NPs

In addition to their utilization as templates, nanostructured modifications of PS microspheres themselves are also desired because, in general, these have fascinating fluorescent, magnetic, catalytic, and biological properties (*27*, *28*). The incorporation of different NPs into one colloidal microsphere, e.g., the so-called Janus particle, is highly desired. Janus particles are named after the two-faced Roman God Janus because these particles may be said to have "two faces" in the form of two distinct types of properties (*29*). These unique surfaces of Janus NPs allow two different types of chemistry to occur on the same particle. Janus particles have unique properties and have potential applications in the fields of SA, colloidal crystals, and sensors (*30–32*).

The generally used preparative methods for Janus particles of PS microspheres include eletrophoretic deposition, Pickering emulsions, and metal evaporation (*33*, *34*). However, all of these strategies suffer from the demerits of being time-consuming, uneconomical, or low yielding. More advanced strategies are still needed for the preparation of anisotropic PS microspheres.

In this part, an attempt to prepare Janus-like PS microspheres in a time-saving and inexpensive way is introduced. 3D SAs were prepared on PS microspheres. Interestingly, the filtration effect was found to play a crucial role in the SA process and the corresponding NP depositions were site-selective. Thus, anisotropic modifications of PS microspheres were accomplished. The mechanism is illustrated in Figure 12 (*35*).

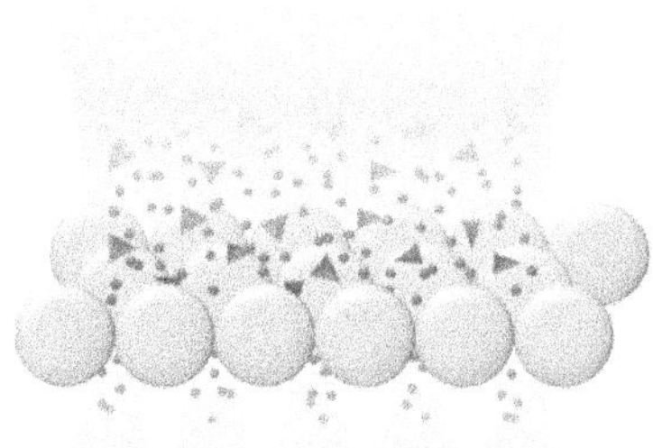

Figure 12. Schematic illustration of the PS microsphere monolayer arrays acting as a network to filter NPs with different sizes. (Reproduced with permission from reference (35). Copyright 2015 Springer).

Figures 13a and b show the SEM images of PS microsphere films after the deposition of big triangular Ag NPs on the top surfaces. The filtration effect for big NPs was confirmed by this step. Small-sized Au NPs were used to further modify the PS microspheres. After the deposition of the Ag NPs, the PDDA on the top surfaces was neutralized by the Ag NPs; therefore, the only possible deposition sites for the Au NPs remained in the lower part of the PS microspheres. Thus, the Au NPs were successively deposited on the surfaces of the lower parts of the PS microspheres. Figures 13c and d show the section-view SEM images of the Ag–Au compound modified PS microspheres. In this way, Janus modification of the PS microspheres was realized in an inexpensive and effective way. The obtained Janus particles show a good *EF* of about 10^6.This is an interesting filtration effect at the nanoscale, which induces the site-selective deposition of NPs on PS microspheres (*35*).

Controlled Ag and Au NP Aggregate Patterns and SERS Application

To investigate the coupling-related SERS performance of NPs, three types of nanostructures were employed, including discrete type, necklace-type (Figure 7), and huddle type (Figure 6b). The detailed synthesis of these substrates is described in a previous paper (*36*). Herein, we focus on the investigation of Raman enhancement and the NP coupling effect related to SERS.

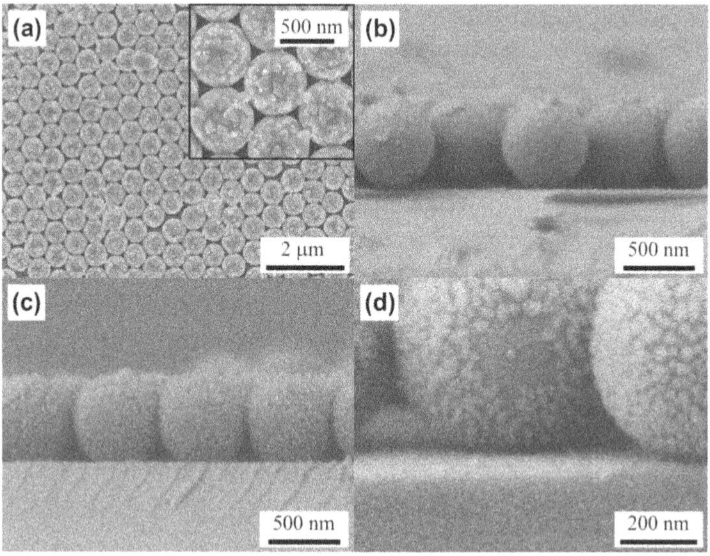

Figure 13. (a) SEM images of PS microsphere monolayers coated with triangular Ag NPs. The inset is the magnified image. (b) SEM section-view of the PS microsphere arrays partially coated with Ag NPs before the deposition of Au NPs. (c) SEM section-view of the PS microsphere arrays partially coated with Ag NPs after the deposition of Au NPs. (d) Magnified SEM image of (c). (Reproduced with permission from reference (35). Copyright 2015 Springer).

Figure 14 shows a cartoon illustration of the three types of samples with Ag NP aggregates in (a) discrete, (b) necklace, and (c) huddle patterns. The differences in the samples are the geometric arrangements of the Ag NPs, i.e., the aggregate patterns. In discrete-type samples, the Ag NPs are arranged almost separately while the Ag NPs are arranged in special ways, including the circular and round patterns in the necklace-type and huddle-type samples. The underlying Au NP films were used as precursor films, guiding the formation of the final aggregate patterns of Ag NPs.

The SERS property was studied on these three types of Ag–Au compound nanostructure substrates by using R6G as the probing molecule. From the obtained SERS spectra, the detection limits were confirmed as 10^{-8}, 10^{-11}, and 10^{-9} mol/L for the discrete-, necklace-, and huddle-type samples, respectively. It is very interesting that NPs with identical species but different aggregate patterns showed obvious differences with regard to the limits of concentrations that can be detected. To assess the differences in the enhancement quantitatively, a statistical method, the so-called ensemble-average method, was employed to uncover the differences in enhancement on the different types of SERS substrates.

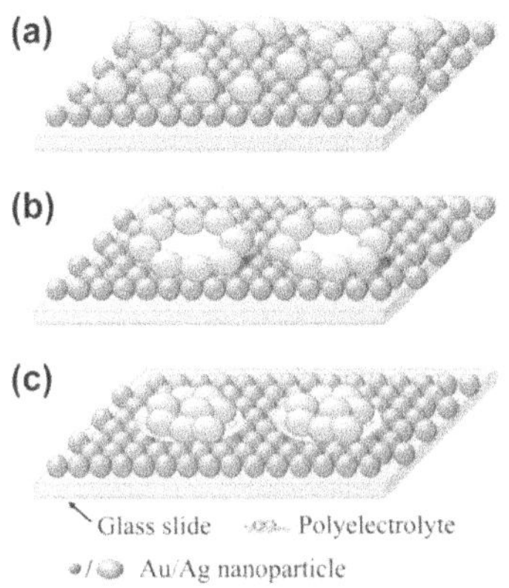

Figure 14. Cartooned illustration of the Ag–Au compound nanostructures with controllable Ag NP aggregate patterns in (a) discrete-, (b) necklace-, and (c) huddle-type patterns. (Reproduced with permission from reference (36). Copyright 2015 John Wiley and Sons).

The SERS *EF* is often used to assess the quality of the SERS substrate. Herein, PS template based pattern (discrete-, necklace-, and huddle-type) from Figure 14 as the representive to evaluate the *EF* was employed. The equation for the calculation of the *EF* is as follows:

$$EF = \frac{I_{SERS} / N_{SERS}}{I_{Raman} / N_{Raman}}$$

where I_{SERS} and I_{Raman} are the spectral intensities of the SERS and normal Raman signals for a selected vibrational mode, and N_{SERS} and N_{Raman} are the numbers of the probing molecules that are excited by the incident Raman laser in SERS and normal Raman, respectively.

This study is more concerned with the relative ratios of the *EFs*. In the *EF* equation, if the probing molecule and the Raman measurement conditions are selected, the I_{Raman} and N_{Raman} values are fixed. In this case, the *EF* will be proportional to the value of I_{SERS}/N_{SERS}. In these experiments, the *EF* was assessed for SERS measurements at a given concentration of 10^{-8} mol/L, which means that N_{SERS} was the same for the discrete-type, necklace-type, and huddle-type substrates. For these three types of substrates, the intensity ratio from the SERS spectra was found to be 1:36.5:30. Thus, the relative ratio of the *EF* for the discrete-, necklace-, and huddle-type substrates (EF_1:EF_2:EF_3) was found to be 1:36.5:30 in a collective way (Figure 14).

The ensemble-average method was employed to further discover the intrinsic physical rules for the enhancement contributions from each plasmonic unit. As counted per unit area of SEM and AFM images of the Ag NPs for the discrete-, necklace- and huddle- type samples, this ratio was determined to be 232:338:387. To assess the ensemble-averaged enhancement contribution of each Ag NP located in different aggregate patterns, the former obtained *EF* ratio was divided by the number of Ag NPs. The final relative, ensemble-averaged *EF* ratio for each Ag NP (*EF'*) located in different aggregate patterns is 1:25:18 (EF_1':EF_2':EF_3'). This result shows that the Ag NPs made noticeably different enhancement contributions to SERS when located in different patterns. The *EF'* of an Ag NP in a necklace pattern is 25 times that of an Ag NP in a single NP pattern, and approximately 1.4 times that of an Ag NP in random pattern. The quantitative study of the contribution of each plasmonic unit is very meaningful for the mechanism of SERS and has the potential to guide the fabrication of new substrates with high SERS performances.

Ultraviolet–visible absorption spectroscopy was used to investigate the intrinsic differences of the plasmonic property of the Ag NPs in different types. The results showed that there was intense coupling in the wavelength range from 400 to 600 nm for the necklace-type samples. This may have contributed to the high *EF*s observed in the SERS spectra for necklace-type samples. These data show that if the arrangement patterns of a certain amount of plasmonic materials are tuned in a controlled manner, then the plasmonic couplings could change correspondingly, inducing obvious differences in the final SERS enhancement. These conclusions are very meaningful for designing and fabricating new types of SERS substrates with high spectral enhancement performance.

All the above-mentioned substrates, which may be fabricated by different techniques, showed difference more or less on the performance of stability, homogeneity, bio-compatibility and sensitivity. Whatever, the invented nanostructures enriched the arsenal of SERS substrates and benefit for both fundamental research and practical applications.

Concluding Remarks

Accompanied by the dramatic developments of nanoscience and laser technology in recent years, the mechanism and applications of optics on the nano-scale are rapidly emerging and developing. As a typical optical phenomenon, SERS has generated broad interest in analytical chemistry, spectroscopic physics, bioassays, diagnostics, forensics, and so on. Due to the intrinsic nature of SERS, substrates are the root of SERS spectroscopy. The control and fabrication of

SERS substrates is always at center-stage in the SERS community. In this chapter, we reviewed the fabricating methods and resulting nanoscale architectures using PS microsphere monolayer templates. The obtained nanostructures included nanowells, nanodots, multilayers, and nanoring arrays of Au, Ag, or Ag–Au compound nanostructures. The SERS performance was compared using controllable Ag NP aggregate patterns of the discrete, necklace, and huddle types. Different types of substrates exhibited different SERS enhancements, due to the aggregate patterns of the plasmonic units. This information is significant for designing new applicable SERS substrates with high performances. The nano-fabrication of SERS substrates is ongoing and promises to be developed to qualify as a standard of practical usage in medicinal, environmental, bioassay, and analytical chemistry.

Acknowledgments

This paper was supported by the Korean Federation of Science and Technology Societies (KOFST) grant funded by the Korean government (MSIP: Ministry of Science, ICT and Future Planning). This paper was also supported by the Natural Science Foundation of China (Grant Nos. 21103062, 21273091, 21221063, 21327803, and 21411140235). W. Ruan is grateful to the Postdoctoral Science Foundation of China (Grant No. 2014M561286).

References

1. Hua, F.; Shi, J.; Lvov, Y.; Cui, T. *Nano Lett.* **2002**, *2*, 1219–1222.
2. Zhang, G.; Surwade, S. P.; Zhou, F.; Liu, H. *Chem. Soc. Rev.* **2013**, *42*, 2488–2496.
3. Muñoz-Bonilla, A.; Fernández-García, M.; Rodríguez-Hernández, J. *Prog. Polym. Sci.* **2014**, *39*, 510–554.
4. Suh, K. Y.; Park, M. C.; Kim, P. *Adv. Funct. Mater.* **2009**, *19*, 2699–2712.
5. Willets, K. A.; Van Duyne, R. P. *Annu. Rev. Phys. Chem.* **2007**, *58*, 267–297.
6. Vekris, E.; Kitaev, V.; Perovic, D. D.; Aitchison, J. S.; Ozin, G. A. *Adv. Mater.* **2008**, *20*, 1110–1116.
7. Stein, A.; Schroden, R. C. *Curr. Opin. Solid State Mater. Sci.* **2001**, *5*, 553–564.
8. Stein, A.; Li, F.; Denny, N. R. *Chem. Mater.* **2008**, *20*, 649–666.
9. Stein, A. *Microporous Mesoporous Mater.* **2001**, *44*, 227–239.
10. Li, Y.; Zhang, J.; Wang, T.; Zhu, S.; Yu, H.; Fang, L.; Wang, Z.; Cui, L.; Yang, B. *J. Phys. Chem. C* **2010**, *114*, 19908–19912.
11. Schmudde, M.; Grunewald, C.; Goroncy, C.; Noufele, C. N.; Stein, B.; Risse, T.; Graf, C. *ACS Nano* **2016**, *10*, 3525–3535.
12. Li, X.; Wang, T. Q.; Zhang, J. H.; Yan, X.; Zhang, X. M.; Zhu, D. F.; Li, W.; Zhang, X.; Yang, B. *Langmuir* **2010**, *26*, 2930–2936.
13. Ruan, W. D.; Lu, Z. C.; Ji, N.; Wang, C. X.; Zhao, B.; Zhang, J. H. *Chem. Res. Chin. Univ.* **2007**, *23*, 712–714.

14. Ruan, W. D. Two Ways to Fabricate Ordered Thin Films & Their SERS Study. Ph.D. Thesis, Jilin University, June 2007.
15. Ruan, W. D.; Lu, Z. C.; Zhou, T. L.; Zhao, B.; Niu, L. *Anal. Methods* **2010**, *2*, 684–687.
16. Oćwieja, M.; Adamczyk, Z.; Morga, M.; Michna, A. *J. Colloid Interf. Sci.* **2011**, *364*, 39–48.
17. Ottakam Thotiyl, M. M.; Basit, H.; Sánchez, J. A.; Goyer, C.; Coche-Guerente, L.; Dumy, P.; Sampath, S.; Labbé, P.; Moutet, J.-C. *J. Colloid Interf. Sci.* **2012**, *383*, 130–139.
18. Andres, C. M.; Kotov, N. A. *J. Am. Chem. Soc.* **2010**, *132*, 14496–14502.
19. Ngo, Y. H.; Li, D.; Simon, G. P.; Garnier, G. *J. Colloid Interf. Sci.* **2013**, *405*, 71–77.
20. Ruan, W. D.; Zhou, T. L.; Hui, G.; Wang, Y.; Chong, X. L.; Wang, X.; Song, W.; Han, X. X.; Zhao, B. *J. Colloid Interf. Sci.* **2014**, *432*, 65–69.
21. Ruan, W. D.; Wang, C. X.; Ji, N.; Lu, Z. C.; Zhou, T. L.; Zhao, B.; Lombardi, J. R. *Langmuir* **2008**, *24*, 8417–8420.
22. Zhang, Y.; Gao, W.; Yang, S.; Liu, S.; Zhao, X.; Gao, M.; Wang, Y.; Yang, J. *J. Raman Spectrosc.* **2013**, *44*, 1666–1670.
23. Wang, Y.; Zhao, X.; Gao, W.; Chen, L.; Chen, S.; Wei, M.; Gao, M.; Wang, C.; Zhang, Y.; Yang, J. *RSC Adv.* **2015**, *5*, 7454–7460.
24. Zhang, Y.; Wang, C.; Wang, J.; Chen, L.; Li, J.; Liu, Y.; Zhao, X.; Wang, Y.; Yang, J. *Spectrochim. Acta A* **2016**, *152*, 461–467.
25. Wang, Y.; Zhao, X.; Chen, L.; Chen, S.; Wei, M.; Gao, M.; Zhao, Y.; Wang, C.; Qu, X.; Zhang, Y.; Yang, J. *Langmuir* **2014**, *30*, 15285–15291.
26. Wang, Y.; Zhang, M.; Yan, C.; Chen, L.; Liu, Y.; Li, J.; Zhang, Y.; Yang, J. *Colloids Surf., A* **2016**, *506*, 96–103.
27. Zhang, J.; Ling, L. T.; Wang, C. F.; Chen, S.; Chen, L.; Son, D. Y. *J. Mater. Chem. C* **2014**, *2*, 3610–3616.
28. Zhang, M. J.; Li, X. Q.; Kang, S. Z.; Qin, L. X.; Li, G. D.; Mu, J. *RSC Adv.* **2014**, *4*, 37854–37858.
29. Li, F.; Josephson, D. P.; Stein, A. *Angew. Chem., Int. Ed.* **2011**, *50*, 360–388.
30. Kaewsaneha, C.; Tangboriboonrat, P.; Polpanich, D.; Eissa, M.; Elaissari, A. *ACS Appl. Mater. Inter.* **2013**, *5*, 1857–1869.
31. Walther, A.; Mueller, A. H. E. *Chem. Rev.* **2013**, *113*, 5194–5261.
32. Zhang, X. L.; Niu, C. Y.; Wang, Y. Q.; Zhou, S. M.; Liu, J. *Nanoscale* **2014**, *6*, 12618–12625.
33. Lu, Y.; Xiong, H.; Jiang, X.; Xia, Y.; Prentiss, M.; Whitesides, G. M. *J. Am. Chem. Soc.* **2003**, *125*, 12724–12725.
34. Ning, Y.; Wang, C. Y.; Ngai, T.; Tong, Z. *Langmuir* **2013**, *29*, 5138–5144.
35. Ruan, W. D.; Zhou, T. L.; Sun, C. B.; Tao, Y. C.; Lu, F.; Wang, X.; Zhao, B.; Cui, Y. Q. *J. Nanopart. Res.* **2015**, *17*, 388.
36. Ruan, W. D.; Cui, Y. Q.; Lin, F.; Zhou, T. L.; Hui, G.; Wang, Y.; Chen, Y. F.; Lu, F.; Guo, J.; Zhao, B. *J. Raman Spectrosc.* **2015**, *46*, 1117–1123.

Chapter 4

Truncated Power Law Analysis of Blinking SERS

Yasutaka Kitahama*

Department of Chemistry, School of Science and Technology, Kwansei Gakuin University, Sanda, Hyogo 669-1337, Japan
***E-mail: kitahama@kwansei.ac.jp**

In this chapter, blinking surface-enhanced Raman scattering (SERS), which are observed at a single molecule level, in various conditions are analyzed by a truncated power law. The exponents and truncation times in the power law are used for determining the quantitative indices of the blinking SERS. The authors have investigated the dependence on excitation light intensity and wavelength and the concentration of NaCl, citrate anion, which covers the Ag nanoaggregate, and analyte molecule. The dependence of the power law exponent is affected by the one- or two-dimensional random walk of the analyte through the efficiency of the localized surface plasmon resonance (LSPR) and adsorption on the Ag surface. Conversely, the truncation time does not depend on the efficiency of the LSPR, but on the excitation wavelength. It suggests that truncation originates from a periodic optical trapping potential well owing to the coupling of multipolar resonances. The dependence of the truncation time on sample conditions, NaCl, citrate anion, and analyte, can estimate the speed of the molecular random walk on the Ag surface, except at the junction. It can hardly be detected even by super-resolution SERS imaging.

Introduction

The incident light on noble metal nanoaggregates is resonant with plasmons, which are dipolar oscillations of conduction band electrons. This is called localized surface plasmon resonance (LSPR). Then, the electromagnetic (EM) field is enhanced at a junction of the nanoaggregate. The surface-plasmon-enhanced EM field remarkably (10^{12}-fold) enhances the effective Raman scattering cross-section of molecules adsorbed on a noble metal nanoaggregate (*1–7*). This phenomenon is called surface-enhanced Raman scattering (SERS). SERS is sufficiently sensitive to measure the Raman spectrum of a single molecule adsorbed on the nanometer-sized junction of a nanoaggregate (*1–60*). Thus, a very small amount of analyte can be identified by SERS owing to the corresponding vibrational fingerprints. Furthermore, the distribution of analytes has been visualized sensitively by inserting nanoporous Ag microparticles and tracking the SERS-active nanoparticles (*6*, *61*, *62*), in a similar manner to semiconductor quantum dots (QDs) (*63–68*).

At a single molecule level, however, the blinking of the SERS is observed. The peaks of the blinking SERS disappear, fluctuate, and slightly shift from the corresponding Raman peaks. Because these phenomena disturb the reproducibility and signal-to-noise ratio of the spectra (*6–40*), it becomes difficult to identify the analyte and investigate its molecular structure. The blinking of the SERS is probably induced by a single molecule that enters and leaves an enhanced EM field at a nanometer-sized junction (a 'hot site') (*4–12*). Therefore, blinking is considered as evidence for single molecule detection similar to the intensity of SERS that shows a Poisson distribution (*13*) and bi-analyte technique, in which unique vibrational signatures are observed from mixed analytes (*48–53*). However, there is a possibility that the blinking SERS is due to a fluctuation of the enhanced EM field at the junction of an Ag nanoaggregate (*17*, *41*). The origin of the fluctuation may be the change of the LSPR by alternation of the junction size and/or refractive index of the surrounding medium. Moreover, SERS of amorphous carbon, which are formed by photo- and thermo-degradation of the adsorbed molecules, show similar behavior (*54–60*). Thus, the mechanism of blinking SERS is still controversial.

Blinking SERS and single molecule SERS have been studied considerably (*7–60*), although only a few quantitative investigations on the dependence of blinking SERS have been performed (*16–24*, *42*). One of them is an investigation using the autocorrelation function, which indicates some periodicity (*15–17*). It has also been applied to fluorescence correlation spectroscopy, which can measure the diffusion coefficient and concentration of molecules that enter and leave a focal area (*69*, *70*). However, the autocorrelation function was not reproduced by any simple function, suggesting a complex process in the blinking SERS (*17*). In another quantitative analysis (*42*), the normalized standard deviation score was derived from the temporal trajectories of the SERS with a resonance Raman effect, namely, the surface-enhanced resonance Raman scattering (SERRS) and surface-enhanced fluorescence (SEF) intensities. A large score means instability in the total intensity of SERRS and SEF. The instability is inversely proportional to the EM enhancement factor of the nonradiative process. Thus, a molecule whose energy transfer to a metal hardly ever takes place shows active blinking.

The blinking SERS from a single Ag nanoaggregate with an adsorbed dye molecule can be analyzed by a power law (*7, 19–24*), in a manner similar to that employed in studies of blinking fluorescence from a single QD (*63–68*). Power law statistics have been widely applied to the analysis of long-range ordered non-exponential behaviors, especially 'self-similarity (fractal)', where features of objects are considered to be similar if the length scales are expanded (*71*). In the blinking SERS, the durations of the bright and dark events range from several tens of milliseconds, which may be a mere limit on the time resolution, to several tens of seconds, and thus a power law is a convenient analysis method for the blinking SERS. The power law in the blinking statistics of the fluorescence from a QD can be explained by the distribution of the first time passage required for a random walker to return to its starting point, for the random walk of a photo-excited electron in a QD (*63*). They may be similar to the blinking statistics of SERS due to the random walk of an adsorbed molecule on a noble metal nanoaggregate. To the best of our knowledge, an analysis of the blinking SERS by using a power law is still limited (*7, 19–24*).

Blinking SERS have been analyzed using a power law with an exponential function, namely, a truncated power law (*7, 20–24*). The truncated power law analysis was compared with the power law analysis for blinking fluorescence from a single QD and other analysis for blinking SERS (*7*). In the power law for blinking SERS, the probability distributions of dark and bright SERS events against their duration are reproduced by a power law with and without an exponential function, respectively. In contrast, those of dark and bright fluorescence events against their duration are reproduced by a power law without and with an exponential function, respectively (*63–68*). The truncation times are independent of the LSPR wavelength and the intensity of excitation laser, although those for a single QD are not affected by the excitation wavelength, but by the excitation laser intensity (*64–66*). By using averages of durations of dark SERS events like those of bright SERS events (*7, 17, 21*), the behavior of molecules in dark SERS states cannot be investigated. The reason is that the averages cannot be derived; namely, total durations of dark SERS events on the single Ag nanoaggregates are decreased by an increase in the number of the events, while total durations of bright SERS events are increased (*7*).

In this chapter, the blinking SERS in various conditions of the excitation light and sample are analyzed by the truncated power law (*21–24*). The power law exponent can be derived from the dimension of the molecular random walk on the Ag surface. The dimension depends on the enhanced EM field (which is influenced by the intensity and wavelength of excitation light and the concentration of NaCl) and the adsorption on the Ag surface (which is affected by the concentration of citrate anion). On the other hand, the truncation in the power law is induced by the periodicity of the enhanced EM field (which does not depend on the intensity, but on the wavelength of the excitation light) and the fast random walk of molecules adsorbed on the Ag surface, which is affected by the concentration of NaCl, citrate anions, and dye molecules and electrostatic attraction between the Ag surface and ionic dye. These are quite significant both in terms of the exploration of the essence, as well as the suppression of blinking for possible applications to single molecule SERS.

Materials and Methods

The details of this experimental setup and procedure have been described elsewhere (*20–24*). In brief, the citrate-reduced colloidal Ag nanoparticles were prepared using the Lee-Meisel method (*72*). 3,3′-Diethyl-5,5′-dichloro-9-methylthiacarbocyanine iodine salt (TCC), 3,3′-diethylthiacyanine iodine salt (cationic TC), or 3,3′-disulfopropylthiacyanine triethylamine (zwitterionic TC), which has one $>N^{+}=$ and two $-SO_{3}^{-}$, were used as analyte dye molecules. A stock aqueous solution of the cationic or zwitterionic dye (2, 4, or 10, 25 μM, respectively), a NaCl aqueous solution (1, 5, or 10, 100 mM, respectively), and an Ag colloidal suspension (72 pM with an average diameter of 70 nm estimated by the extinction spectrum and Mie theory) were mixed in a volume ratio of 1 : 1 : 2. The sample solution was spin-coated onto a glass plate. An aliquot of a 1 M NaCl solution was dropped on the glass plate to immobilize the sample Ag nanoaggregates on the surface. The glass plate was covered with another glass plate to prevent the solution from evaporating.

The Ag nanoaggregates with adsorbed dye molecules were excited using the unfocused beam of the 458, 514, or 568 nm line of an Ar or Kr ion laser with an intensity of 8–20 mW, which corresponds to a power density of 80–200 W cm^{-2}. Videos of the blinking SERS were taken for 20 min by an inverted microscope (Olympus, IX-70) coupled with a cooled digital CCD camera (Hamamatsu, ORCA-AG) whose time-resolution was set at ~60 ms.

Truncated Power Law Analysis

Figure 1a shows video microscope images of blinking SERS from two Ag nanoaggregates (see the upper left and the lower right) with adsorbed cationic TC excited at 458 nm (*22*). It is noted that a bright spot in the upper left of the fourth and fifth video images vanishes in the sixth image and then emits light again in the 37th image at the same position during the excitation.

Figure 1b exhibits temporally fluctuated SERS spectra from single Ag nanoaggregates with adsorbed TCC excited at 568 nm (*22*). These prominent peaks were observed at almost the same wavenumbers as the peaks in a conventional Raman spectrum of the dye powder except for the peak at 920 cm^{-1}. The disappearance, fluctuation, and slight shift of the SERS peaks from the corresponding Raman peaks have already been reported (*7–40*). For several analytes, the spectral fluctuation is correlated with vibronic coupling; nontotally symmetric modes show a sharper blinking behavior than totally symmetric modes (*26*). In the case of metal-free tetraphenylporphine, the fluctuated SERS peaks originate from a structural change of the molecule (*27*). When the adsorbates are degraded photochemically and thermally by the enhanced EM field, sharp peaks of amorphous carbon are observed at random positions that average out two broad maxima at approximately 1300 and 1550 cm^{-1} (*54–59*). Furthermore, the number of SERS peaks of the dye decreases to two lines at approximately 1350 and 1600 cm^{-1}, which are assigned to the D and G modes of sp^2 carbon clusters, respectively, while the dyes are irradiated by the laser (*60*).

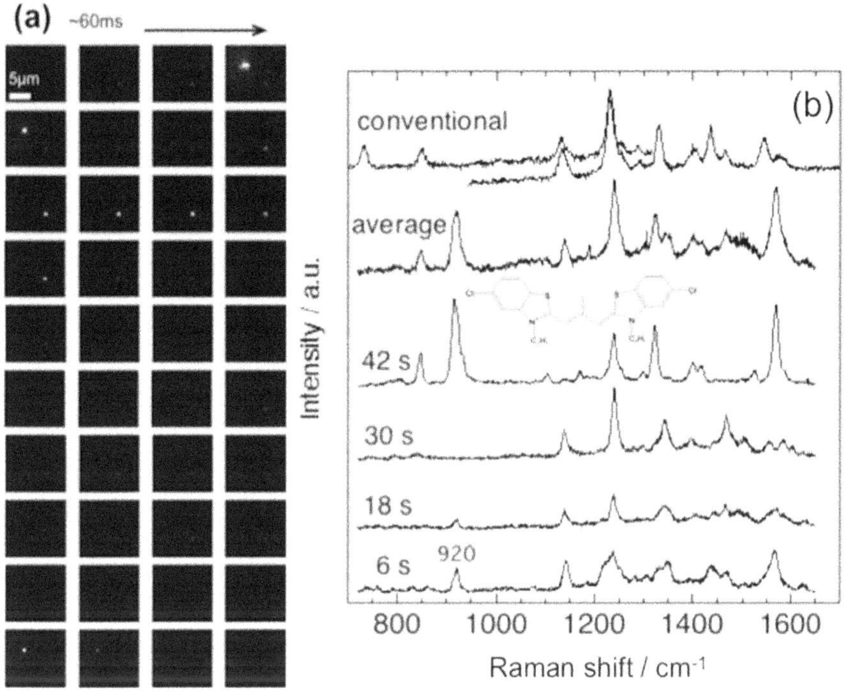

Figure 1. (a) Video microscope images and (b) temporally fluctuated spectra of SERS from a single Ag nanoaggregate with adsorbed cationic TC and TCC excited at 458 and 568 nm, respectively. In (a), the scale bar is 5 μm and the exposure time per image is ~60 ms. In (b), the integration time per spectrum is 6 s. The top panel in (b) shows a conventional Raman spectrum of TCC excited at 785 nm. Insets in (b) show the molecular structure of TCC. (Reproduced with permission from ref. (22). Copyright 2011 Royal Chemical Society.)

In the spectra shown in Figure 1b, however, it is thought that the heat of the molecule adsorbed on the junction because of the enhanced EM field is quickly transferred to the Ag nanoaggregate (*73*, *74*). Indeed, the anti-Stokes to Stokes intensity ratios for the SERS peaks are attributed not to a Boltzmann distribution at high temperatures, but to the selective enhancement of the SERS peaks that are close to the LSPR band (*75*). Thus, these indicate that the blinking may be attributed not to the SERS of amorphous carbon, but to the SERS of the dye molecules. At low concentrations of the adsorbate, the molecules barely emit fluorescence owing to the energy transfer to the metal because they stay in close proximity to a metallic surface (*75*). Thus, the blinking emission of the dye adsorbed on a single Ag nanoaggregate may be mainly attributed not to fluorescence, but to SERS. Moreover, it has been reported that SERS signals are strongly related to the background emission. As a result, the SERS blinking intensity can be represented by a time-profile of the emission intensity from a single Ag nanoaggregate with the adsorbed molecule (*17*, *76*).

Figure 2a displays the time profile of the blinking SERS intensities of TC adsorbed on single Ag nanoaggregates (*24*). From the baseline, an averaged intensity I_{base} and the standard deviation σ were evaluated. Bright and dark SERS events are defined as the events showing larger and smaller intensities than a threshold of $I_{base} + 3\sigma$. The probability distribution for a duration t_1 of the bright or dark SERS events is given by

$$P(t_1) = \sum_{t=t_1} n(t)/t_1, \tag{1}$$

where $n(t)$ is the number of bright or dark SERS events against each duration. Equation (1) represents the number of events for durations $\geq t_1$ that are summed up, and then the summation divided by the duration t_1 equals the probability distribution. Figure 2b shows the probability distributions of the bright and dark SERS events, respectively, against their durations (*24*). In the bright SERS, the logarithm-logarithm (log-log) plot shows the line given by a power law as

$$P_{on}(t) = A_{on}t^{\alpha_{on}} \tag{2}$$

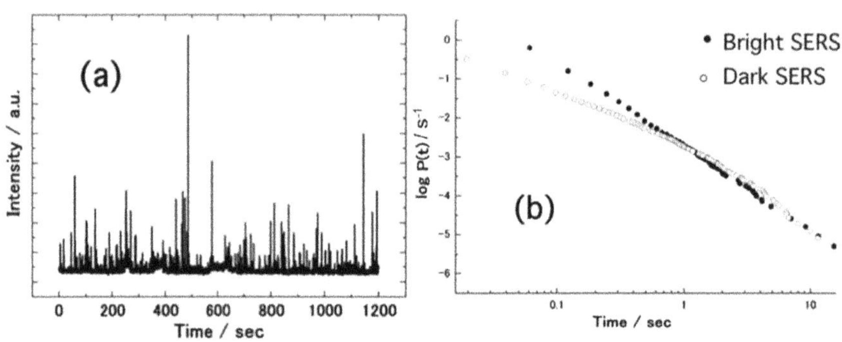

Figure 2. (a) Time profiles of the 458 nm-excited SERS intensities of zwitterionic TC adsorbed on single Ag nanoaggregates. (b) The probability distributions of the bright and dark SERS events against their durations. (Reproduced with permission from ref. (24). Copyright 2015 Royal Chemical Society).

In the dark SERS, the log-log plots were truncated at their tails and were reproduced by a power law with an exponential function as

$$P_{off}(t) = A_{off}t^{\alpha_{off}}\exp(-t/\tau) \tag{3}$$

where $P_{on,off}(t)$ are the probability distributions of the bright and dark SERS events in each case, respectively; $A_{on,off}$ are the coefficients for normalization; $\alpha_{on,off}$ are the power law exponents in the bright and dark SERS, respectively; and τ is the truncation time in the power law.

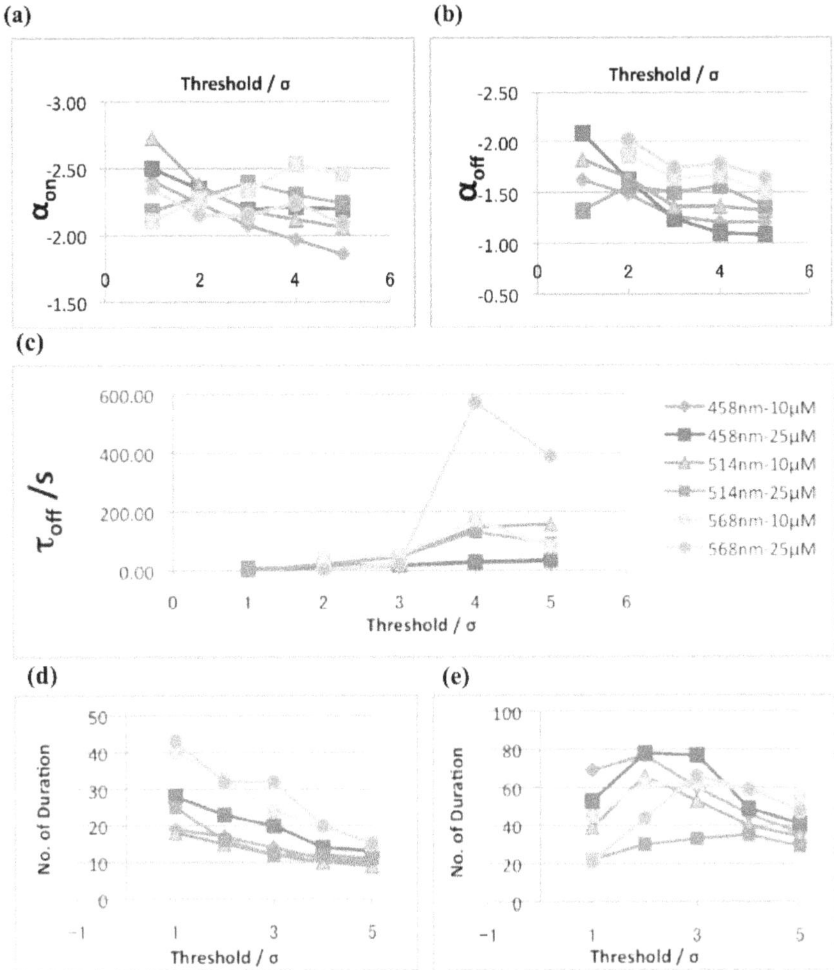

Figure 3. Power law exponents for (a) bright and (b) dark SERS events, (c) the truncation times in the fitting for the probability distributions against durations, and the number of durations in (d) bright and (e) dark SERS events from single Ag nanoaggregates with adsorbed zwitterionic TC with 100 mM NaCl by using the various thresholds for the definition of the bright and dark events. (Reproduced with permission from ref. (23). Copyright 2013 American Chemical Society).

The influence of the threshold for the definition of bright and dark SERS events on the blinking analysis has been checked (*23*). Figures 3a–3c show the exponents and the truncation times derived by using the different thresholds. Most of the exponents (α_{on} and α_{off}) and truncation times under different conditions increase with the threshold (in Figures 3a and 3b, the y-axes are upside down) (*23*). The reason may be as follows. The bright SERS event with the longest duration is divided into a few events for long durations rather than many events for short durations by the higher threshold. In other words, the rightmost point in the log-log plot (like Figure 2b) vanishes, and the neighbor points go upward. Conversely, the probabilities of dark SERS events for long durations are increased by the higher threshold; namely, the points on the lower right in the log-log plot (like Figure 2b) go upward. Thus, the exponents and truncation time are equally increased. When these indices are different in the trends, the dependence of blinking SERS on other indices can be discovered. Figures 3d and 3e represent the number of durations (the points in the log–log plot) for the bright and dark SERS events, which are decreased and almost constant (*23*), respectively, by using a higher threshold because bright SERS events for long durations are rarely counted.

In Figure 4, the top and middle panels show m_{on} ($-m_{on}$ is the power law exponent and corresponds to $\alpha_{on,off}$ in this chapter) and the truncation time for the bright fluorescence event, respectively, from a single QD whose fluorescence intensity histograms did not show two distinct peaks. These were derived from the different thresholds, which are represented by the x-axis (*67*). The value of m_{on} increases at the beginning and then becomes almost stable, and the truncation times decrease as the threshold is increased. For the blinking SERS, on the other hand, both $-\alpha_{on}$ and $-\alpha_{off}$ (which corresponds to m_{on} in Figure 4) are decreased as shown in Figures 3a and 3b (the y-axes are α_{on} and α_{off}, but are upside down), and the truncation time is increased using the higher threshold as exhibited in Figure 3c (*23*). The reason for the opposite trend of the SERS blinking is likely that truncation describes an event that is opposite from fluorescence blinking.

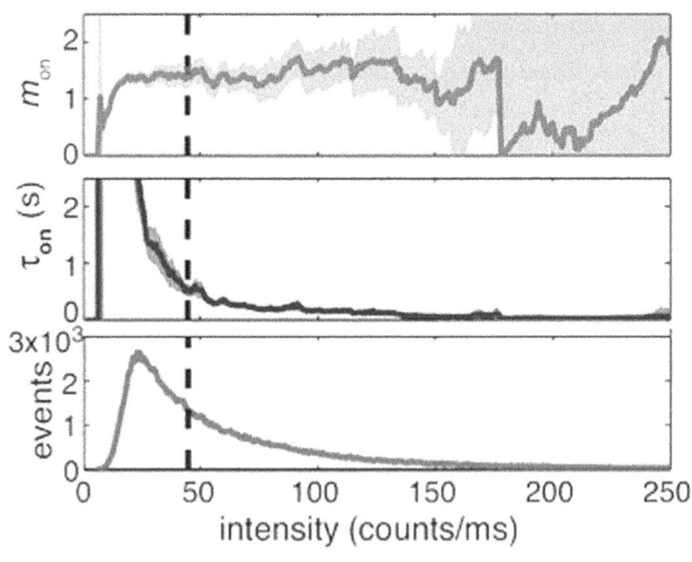

Figure 4. Exponents (top) and truncation times (middle) in the power law against the threshold, and the number of the bright fluorescence events (bottom) against the intensity from a single QD whose fluorescence shows various intensities like the blinking SERS. (Reproduced with permission from ref. (67). Copyright 2010 American Chemical Society).

Power Law Exponents for the Bright and Dark Events

Physical Interpretation of the Power Law Exponents

Figures 5a and 5b show time-profiles of blinking SERS intensities from cationic TC adsorbed on single Ag nanoaggregates whose LSPR bands appear at ~460 and ~590 nm, respectively, excited at 458 nm (*22*). The single Ag nanoaggregate with the LSPR band at ~460 nm displayed a bumpy baseline in the time-profile, as shown in Figure 5a. For example, a bright SERS event took place for ~200, 500, and 750 s. Conversely, the baseline from the nanoaggregate with the LSPR band at ~590 nm was almost flat, as shown in Figure 5b. The characteristics of these time-profiles can be quantified by a power law. For the bright SERS, Figure 5c shows that the line corresponding to Figure 5a is sloped more gently than that corresponding to Figure 5b (*22*). Their power law exponents are also derived: $\alpha_{on} = -2.24$ and -2.72 correspond to Figure 5a and 5b, respectively. These two exponents represent that the probabilities of a long duration in the bright SERS by excitation at the LSPR wavelength are higher than those by excitation at a wavelength that is different from the LSPR wavelength. In contrast, Figure 5d indicates that the slopes of the lines and exponents in the dark SERS show opposite trends ($\alpha_{off} = -1.42$ and -1.12 correspond to Figures 5a and 5b, respectively) to those for the bright SERS (*22*).

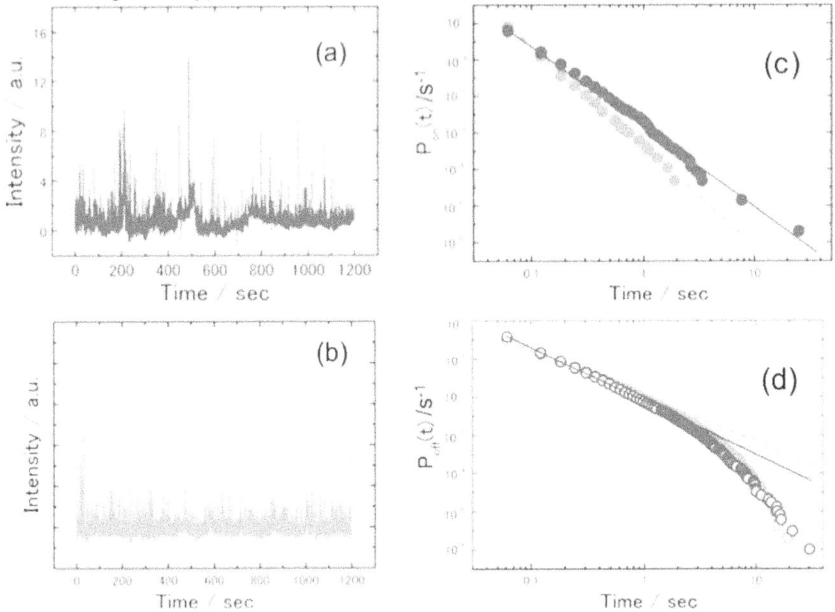

Figure 5. Time-profiles of 458 nm-excited SERS intensities of cationic TC adsorbed on single Ag nanoaggregates with LSPR at (a) 460 and (b) 590 nm (19). The probability distributions of the 458 nm-excited (c) bright and (d) dark SERS events against durations for TC adsorbed on single Ag nanoaggregates with LSPR at 460 nm (blue (black in the print version) circles and solid lines) and 590 nm (orange (gray in the print version) circles and broken lines). (Reproduced with permission from ref. (22). Copyright 2011 Royal Chemical Society).

Figure 6 schematically illustrates the fact that the duration of the dark or bright SERS event is related to the efficiency of LSPR, which induces the surface-plasmon-enhanced optical potential well around the junction ('hot site'), by analogy with the case of a single QD (*77*). At the junction of the Ag nanoaggregate, a molecule can barely escape from the deeper optical trapping potential well under the LSPR condition; namely, the LSPR band of the Ag nanoaggregate appears near the excitation wavelength. The probability of a long duration in the bright SERS is increased by the deeper 'hot site'. Thus, the lines given by the log-log plots for the bright SERS may be sloped more gently, and the α_{on} probably increase as the optical potential well becomes deeper. In contrast, a molecule may stay shorter on the Ag surface except for the junction, where no SERS is emitted, with a broadening of the 'hot site' and a narrowing of the rest of the surface. It is likely that the probability of a long duration in a dark SERS decreases as the 'hot site' broadens. Thus, the lines given by the log-log plots for the dark SERS may be sloped more steeply, and the α_{off} probably decreases with a broadening of the optical potential well.

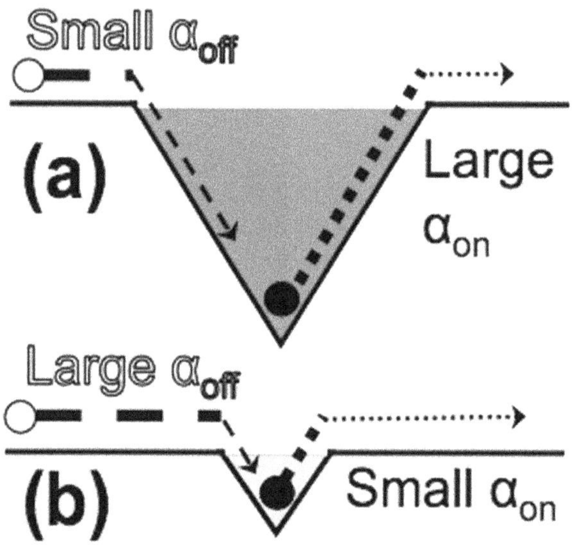

Figure 6. Schematic illustrations of molecular movement around the junction of the Ag nanoaggregate on the condition of (a) efficient and (b) inefficient LSPR.

Figures 7a and 7b show the power law exponents for the bright and dark SERS events from the Ag nanoaggregates whose LSPR bands appear at various wavelengths (*22*). As the LSPR wavelength approaches the excitation wavelength, the power law exponents in the bright and dark SERS increase and decrease, respectively. In other words, the probabilities of a long duration in the bright SERS and a short duration in the dark SERS increase as a result of the resonance of the plasmon with the excitation light. The calculated depths and widths of the optical trapping potential wells at a gap in Ag nanoparticles are obtained at various excitation wavelengths as exhibited in Figures 7c and 7d, respectively (*22*). The calculated LSPR band of the nanodimer appears at 455 nm. It should be noted that the calculated results (Figures 7c and 7d) are roughly consistent with the experimental results (Figures 7a and 7b). The calculation suggests that the optical potential wells are deepened and broadened by the LSPR wavelength approaching the excitation wavelength, and then the increase of α_{on} and the decrease of α_{off} are induced, respectively.

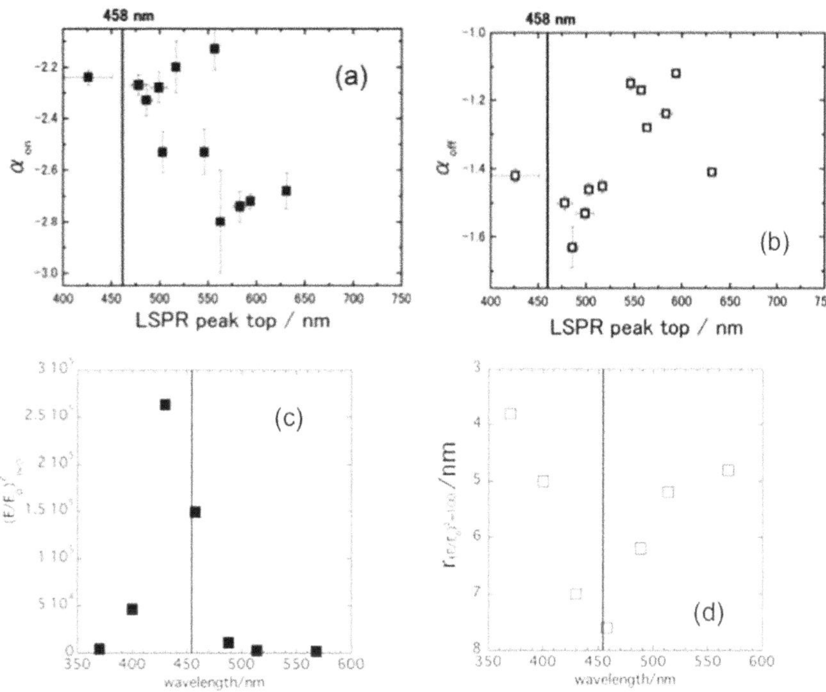

Figure 7. Power law exponents of the fitting for the probability distributions of (a) bright and (b) dark SERS events against duration times at various LSPR wavelengths for single Ag nanoaggregates with adsorbed cationic TC excited at 458 nm. Calculated (c) depths and (d) half-widths where there is a 100-fold enhancement of optical trapping potential wells at the gap (1 nm) in Ag nanoparticles with a diameter of 20 nm, whose LSPR band appears at 455 nm, excited by vertical polarizations at various wavelengths. (Reproduced with permission from ref. (22). Copyright 2011 Royal Chemical Society).

Similar to the study of the blinking fluorescence from a single QD, the power laws can be explained by the distribution of the first-passage time required for the random walker to return to its starting point for a one- or two-dimensional random walk (*63*), which led to the use of Pascal's triangle, as shown in Figure 8, which schematically illustrates the probability to return to the starting point for a one-dimensional random walker at a time. The probabilities for the one- or two-dimensional random walker to return to its starting point for the first time at time 2*n*, where *n* is the number of steps of the random walker, are given as

$$P_{2n} = \binom{2n-2}{n-1}2^{-2n+2} - \binom{2n}{n}2^{-2n} = \frac{1}{2n-1}\binom{2n-1}{n}2^{-2n+1} \quad (4)$$

or

$$P_{2n} = \binom{2n-2}{n-1}^2 4^{-2n+2} - \binom{2n}{n}^2 4^{-2n} \tag{5}$$

which is approximately proportional to $n^{-3/2}$ or n^{-1}, respectively, through Stirling's formula (*78*). Thus, the power law exponents for the one- or two-dimensional random walk model are derived to be −1.5 and −1, respectively. A three-dimensional random walker does not always return to its starting point unlike one- and two-dimensional random walkers, which is called Polya's theorem (*63*, *78*), and thus cannot show continuous blinking.

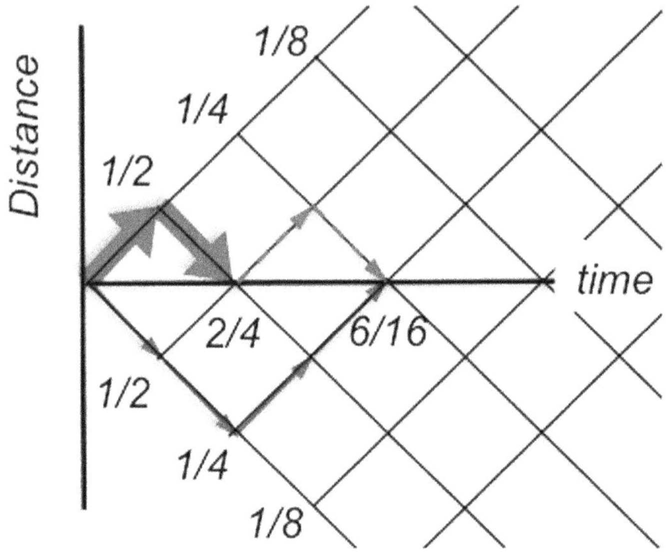

Figure 8. Schematic illustrations of time and probability (italic) to return to the starting point for a one-dimensional random walker.

Effect of Surface-Plasmon-Enhanced Electromagnetic Field

Dependence on Excitation Light Intensity

In the experiment, an aqueous solution of TCC (4 μM) with polyacrylic acid (ratio of the residue to TCC was 2 : 1), a NaCl aqueous solution (5 mM), and an Ag colloidal suspension were mixed in a volume ratio of 1 : 1 : 2 (*22*). For the SERS-active single Ag nanoaggregates, the LSPR bands appear at various wavelengths.

Figures 9a–9b and 9c–9d show the histograms of the power law exponents for the bright and dark SERS events, respectively, under different excitation laser intensities (*22*). As the laser intensity increases (11, 14, and 17 mW correspond to power densities of 110, 140, and 170 W/cm^2, respectively), the maxima in the histograms of the power law exponents for the bright and dark SERS increase from $\alpha_{on} \sim -2.0$ to −1.8 and decrease from $\alpha_{off} \sim -1.3$ to −1.5, respectively.

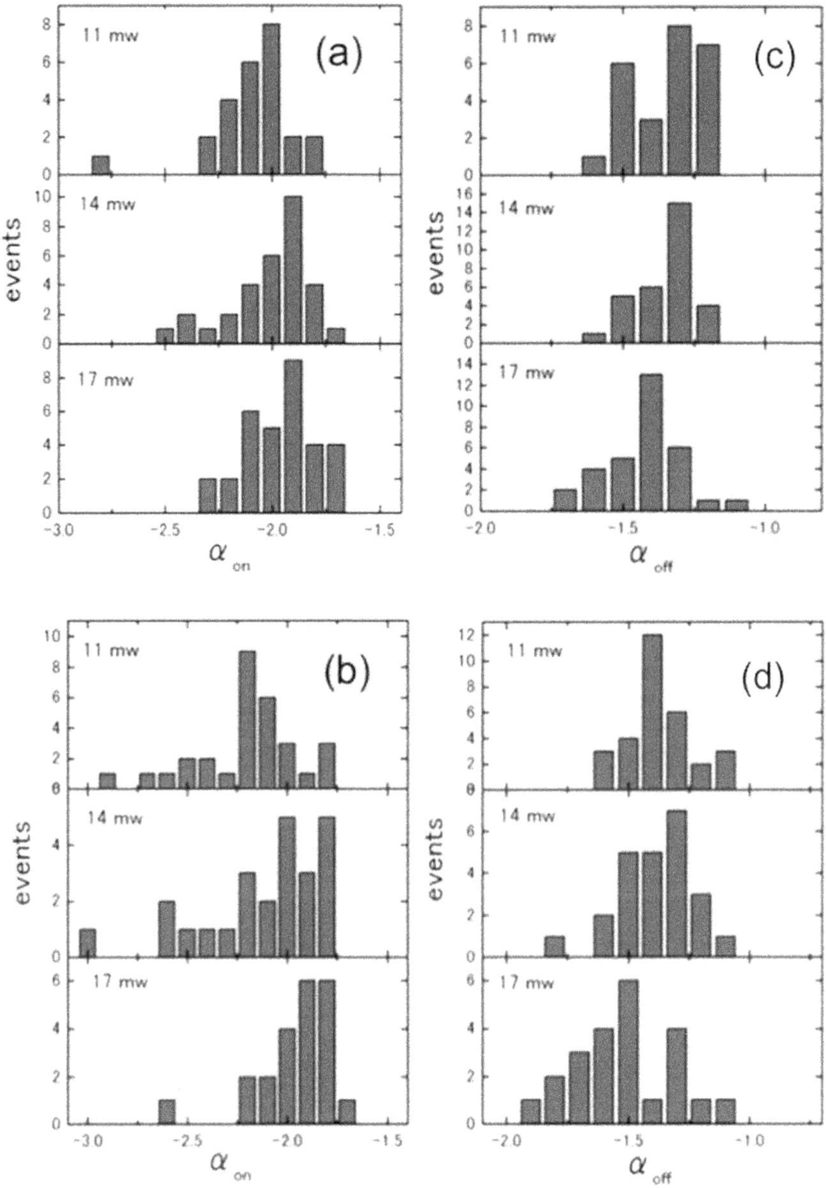

Figure 9. Histograms for the exponents of the power law that reproduce the probability distributions of (a, b) bright and (c, d) dark SERS events by the excitation at (a, c) 514 or (b, d) 568 nm against their durations for TCC adsorbed on single Ag nanoaggregates excited under the laser intensities of 11, 14, and 17 mW corresponding to power densities of ~0.11, 0.14, and 0.17 kW/cm², respectively. (Reproduced with permission from ref. (22). Copyright 2011 Royal Chemical Society).

The maxima in the histograms of the power law exponents for the bright and dark SERS approach –1.5 as the laser intensity increases. For the bright and dark fluorescence events from a single QD, many experimental power law exponents are consistent with –1.5 (*63–66*), which are derived from the one-dimensional random walk model. In the case of a QD, a photo-excited electron in the conduction band, which recombines with a hole in a valence band and then emits fluorescence, is restrained by Coulomb attraction. Therefore, the random walk of the photo-excited electron on a single QD can be approximated as a one-dimensional random walk model as a function of the distance from the hole, rather than a two-dimensional random walk model. This situation may be similar to the SERS-active Ag nanoaggregate; the adsorbate molecule on the surface is restrained by a surface-plasmon-enhanced optical trapping potential well around the junction (*25, 79*). The optical trapping force due to the surface-plasmon-enhanced EM field is strengthened as the original EM field of the excitation light (the laser intensity) increases. Then, the adsorbed molecule is restrained more tightly by the stronger surface-plasmon-enhanced optical trapping force. Thus, the power law exponents near –1.5 for the bright and dark SERS may originate from the one-dimensional random walk in the strong surface-plasmon-enhanced optical trapping force, which is represented by the function as only the distance from the junction.

Under the low excitation laser intensity, the power law exponent deviates from a value of –1.5, indicating other random walk models. The random walk of a molecule adsorbed on an Ag surface that does not have surface-plasmon-enhanced optical trapping potential, where no SERS is emitted, is more likely to be expressed by a two-dimensional random walk model rather than a one-dimensional random walk model. From the two-dimensional random walk model, a power law exponent of –1 is derived (*63*). This result may be consistent with the exponents in the dark SERS under the low laser intensity of 11 mW corresponding to the low power density of 110 W/cm^2 ($\alpha_{off} \sim -1.3$). Also in the dark SERS excited at a wavelength that is different from the LSPR wavelength, the exponents approach $\alpha_{off} \sim -1.1$ as shown in Figure 7b (*22*).

Figure 10 shows semi-logarithmic plots of the probability distribution of dark SERS events from 10^{-11} M Fe-protoporphyrin IX (*19*). The power law exponents for dark SERS events α_{off} were reported to be –1.46, –1.56, and –1.51 (see Figure 10 and its caption, namely $\alpha_{off} = -(1+\alpha)$) like blinking fluorescence from a single QD (*63–66*). The value of ~ –1.5 may be due to the strong laser power density of the focused beam (≤ 1 mW (*19*), which probably corresponds to a power density of ≤ 130 kW/cm^2).

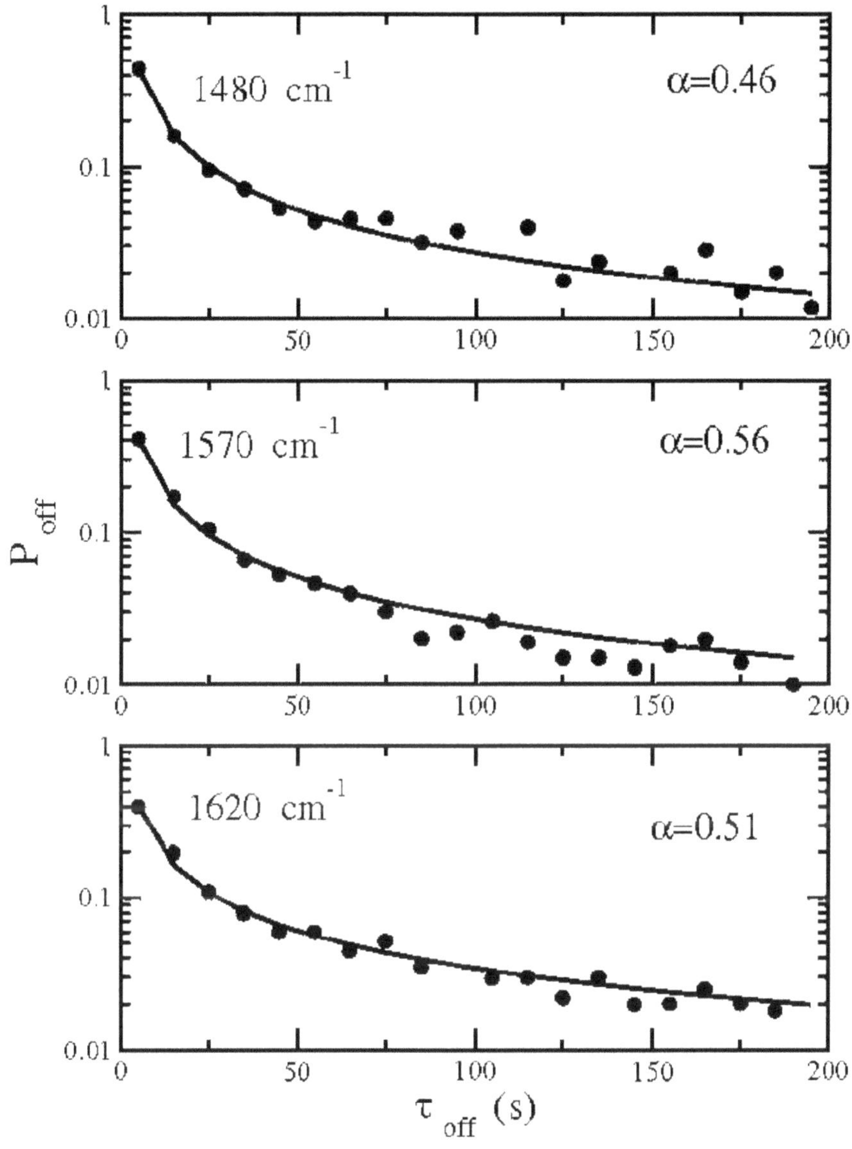

Figure 10. Probability distribution of dark SERS events at the peaks of Fe-protoporphyrin IX (10^{-11} M) adsorbed on an Ag nanoaggregate excited at 514 nm plotted against the duration (τ_{off}). The fitting curves are given by $1/\tau_{off}^{1+\alpha}$. (Reproduced with permission from ref. (19). Copyright 2005 American Physical Society).

Dependence on Excitation Wavelength

In the experiment, an aqueous solution of the twitterionic TC (10 or 25 μM), a NaCl aqueous solution (100 mM), and an Ag colloidal suspension were mixed in a volume ratio of 1 : 1 : 2 (*23*).

Figure 11 shows histograms of the exponents in the power law for the bright and dark SERS events from TC adsorbed on 32–163 single Ag nanoaggregates excited using the 458, 514, or 568 nm line of the laser with the low intensity of 10 mW corresponding to the low power density of 100 W/cm^2 (*23*). Despite the different concentrations of TC (Figures 11a and 11b correspond to 10 and 25 μM TC, respectively), the maxima of the histograms of α_{on} and α_{off} were increased and decreased, respectively, by excitation at longer wavelengths except for α_{off} at 514 nm. The averaged exponents for bright and dark SERS events show a similar trend as summarized in Table 1.

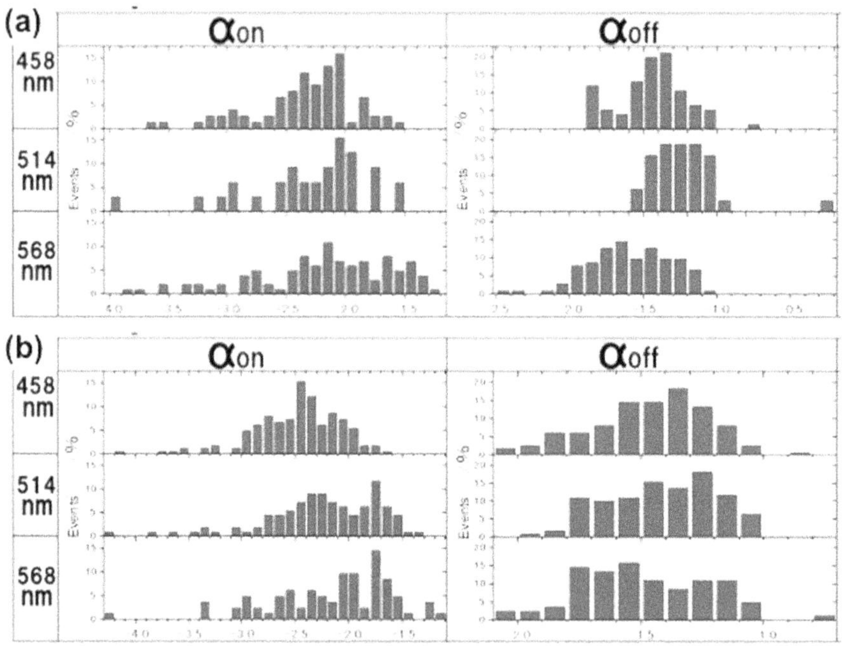

Figure 11. Histograms for exponents of the power law that reproduce the probability distributions of bright and dark SERS events against their durations for (a) 10 or (b) 25 μM zwitterionic TC adsorbed on single Ag nanoaggregates excited at 458, 514, and 568 nm. (Reproduced with permission from ref. (23). Copyright 2013 American Chemical Society).

This trend is similar to the dependence of the power law exponent on excitation light intensity and LSPR wavelength; namely, the α_{on} and α_{off} increased and decreased, respectively, by a deeper surface-plasmon-enhanced optical trapping potential well as the excitation light intensity increased, and the LSPR peaks of the Ag nanoaggregates approached the excitation wavelength (*22*). In this case, a deeper optical trapping potential well may be induced by the excitation at a longer wavelength (especially at 568 nm) through LSPR. It was confirmed that the LSPR peak of the single SERS-active nanodimer mostly appears at 530–750 nm by using a scanning electron microscope (SEM) image, Rayleigh scattering spectrum, and finite-difference time-domain (FDTD) calculation of the same Ag nanodimer (*74*). The LSPR peak of a larger nanoaggregate will appear at longer wavelengths than 530–750 nm, which is the LSPR wavelength of the nanodimer. Thus, the power law exponents may be close to –1.5 by the excitation at a longer wavelength via the approach to the LSPR wavelengths of the SERS-active Ag nanoaggregates.

Table 1. The power law exponents for the bright and dark SERS events (α_{on} and α_{off}) at various excitation wavelengths, at different concentrations of zwitterionic TC (*23*); from the Ag nanoaggregates kept at room temperature and in iced water, which was covered with large and small amounts of citrate anion, respectively (*24*), and at different concentrations of NaCl

α_{on} (R.T.; 100mM)	458 nm	514 nm	568 nm
10 μM	−2.34 ± 0.05	−2.30 ± 0.09	−2.18 ± 0.06
25 μM	−2.48 ± 0.03	−2.24 ± 0.05	−2.16 ± 0.06

α_{on} (458nm; 25μM)	R. T.	Iced water
100 mM	−2.51 ± 0.02	−2.52 ± 0.03
10 mM	−2.19 ± 0.02	

α_{off} (R.T.; 100mM)	458 nm	514 nm	568 nm
10 μM	−1.44 ± 0.03	−1.22 ± 0.04	−1.59 ± 0.03
25 μM	−1.46 ± 0.02	−1.42 ± 0.02	−1.48 ± 0.03

α_{off} (458nm; 25μM)	R. T.	Iced water
100 mM	−1.44 ± 0.02	−1.72 ± 0.02
10 mM	−1.52 ± 0.02	

Dependence on NaCl Concentration

Figure 12 represents the histograms of the power law exponents derived from 181 or 213 single Ag nanoaggregates with twitterionic TC (25 μM) and NaCl (10 or 100 mM, respectively) excited at 458 nm. The maxima for α_{on} and α_{off} increased and decreased, respectively, as the NaCl concentration was decreased from 100 to 10 mM.

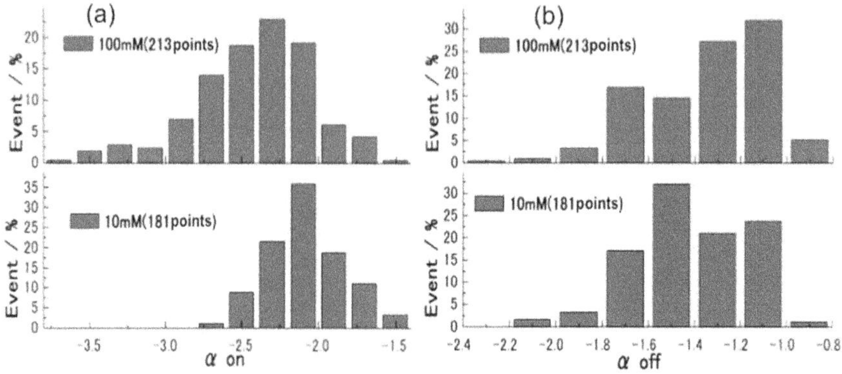

Figure 12. Histograms for exponents of the power law that reproduce the probability distributions of (a) bright and (b) dark SERS events against their durations for zwitterionic TC adsorbed on single Ag nanoaggregates with 100 and 10 mM NaCl.

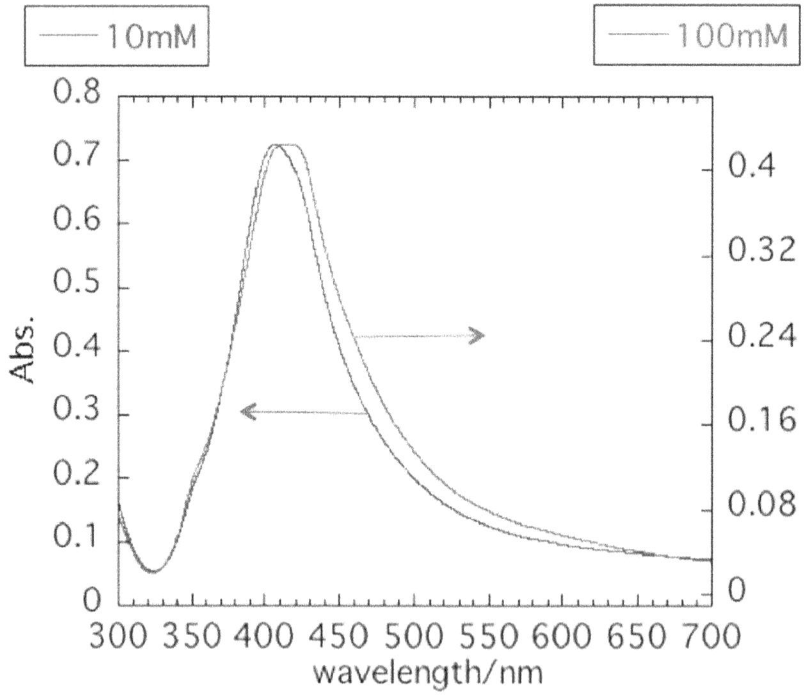

Figure 13. Normalized extinction spectra of the Ag colloidal suspension with 10 and 100 mM NaCl.

Figure 13 shows the extinction band of the Ag colloidal suspension with NaCl. The band at 100 mM NaCl was observed at longer wavelengths than that at 10 mM. It seems that the Ag suspension with 100 mM NaCl is suitable for the LSPR because its absorption band was shifted to 458 nm. However, the LSPR peak of the single SERS-active nanodimer mostly appears at the foot (530–750 nm) of the absorption band of the Ag suspension (*74*). The SERS-active Ag nanodimers grow larger at higher concentrations of NaCl due to salting out. The LSPR peak of the larger Ag nanoaggregate may appear at a wavelength that is different from the excitation wavelength, 458 nm. Thus, the Ag nanoaggregate with NaCl 100 mM likely gives rise to the weaker surface-plasmon-enhanced optical trapping force at the junction, and then the power law exponents for the bright and dark SERS events leave the value of −1.5.

Consequently, the α_{on} and α_{off} were simultaneously increased and decreased, respectively, by a stronger surface-plasmon-enhanced EM field under the more intense excitation light, at the longer excitation wavelength (568 nm), and/or at the lower concentration of NaCl (10 mM). Recently, the SERS-active molecule around the junction has been detected at precise positions by super-resolution imaging (*46*, *47*). It will be interesting to investigate the molecular random walk at the junction of the Ag nanoaggregate, the one- or two-dimensional random walk by using the super-resolution imaging.

Effect of the Adsorption on the Ag Surface

Dependence on Citrate Anion Concentration

The same-sized colloidal Ag nanoparticles that were covered with a large and small amount of citrate anions were used for this experiment. After the heating of the precursor for an hour (*72*), the same suspension was kept in iced water or at room temperature (25°C) overnight (*24*). Extinction spectra of the two types of the Ag colloidal suspension are almost the same (*24*). Thus, the Ag nanoparticles in the two types of suspensions have the same LSPR property because the same suspension was divided after the synthesis by heating for an hour. On the other hand, the zeta potential of the Ag colloidal nanoparticles that were kept at room temperature is more negative (−55.8 ± 2.3 mV) than that in ice water (−47.4 ± 0.2 mV) (*24*). This represents that the former are covered with a larger amount of citrate anions than the latter. The reason is that the citrate anions act as stabilizers of the colloid, which when chemisorbed onto the Ag surface (*80*), are increased at higher temperatures because chemisorption takes place through a transition state. The two types of Ag colloidal suspensions were mixed with an aqueous solution of the twitterionic TC (25 μM) and a NaCl (100 mM) in a volume ratio of 2 : 1 : 1.

Figure 14 shows histograms of the exponents in the power law that reproduce the probability distributions of the bright and dark SERS events against their duration for TC adsorbed on the single Ag nanoaggregates (*24*). The power law exponents were derived from 135 and 213 single Ag nanoaggregates and had been kept in iced water and at room temperature overnight, respectively. For bright SERS events, the maxima of the histograms of the power law exponents, α_{on}, appear at the same value, and the averages are almost the same value within their standard errors as expressed in Table 1. On the other hand, the α_{off} of the Ag nanoaggregates from the colloidal suspension that had been kept at room temperature overnight are smaller than those in iced water as expressed in Figure 14 and Table 1. It is noted that the present result is different from the opposite trend of α_{on} to α_{off} in the above sections; the α_{on} and α_{off} were simultaneously increased and decreased, respectively, by the excitation at the wavelength near the LSPR peak of the Ag nanoaggregate and/or by the stronger excitation light (*22*, *23*).

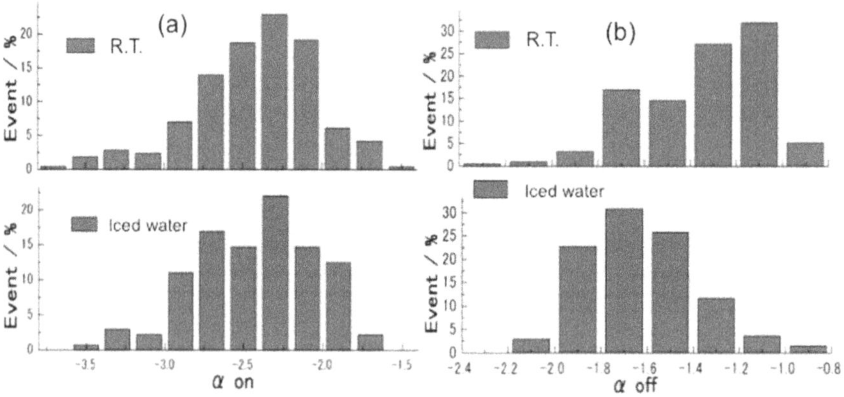

Figure 14. Histograms for exponents of the power law that reproduce the probability distributions of (a) bright and (b) dark SERS events against their durations for zwitterionic TC adsorbed on single Ag nanoaggregates kept at room temperature and in iced water, which are covered with large and small amounts of citrate anions (24).

In this case, the similar α_{on} (Figure 14a) may be induced by the similar surface-plasmon-enhanced optical trapping well through the same LSPR property of the two types of Ag colloidal suspensions due to the same averaged sizes of the nanoparticles (~55 nm (*24*), which was measured by a light scattering photometer).

From the Ag colloidal nanoparticles that were kept at room temperature overnight and in iced water, α_{off} mainly appear at –1.1 and –1.7, respectively (Figure 14b). From the two- and one-dimensional random walk model, the power law exponents are derived to be –1 and –1.5, respectively (*63*). Figure 15 schematically illustrates the molecular random walks on the Ag surface covered with large and small amounts of citrate anions. On the Ag colloidal nanoparticles that were kept at room temperature and then adsorbed a large amount of citrate anions, the zwitterionic TC, which has a one positive-charged nitrogen atom and two -$SO_3{}_-$ atoms, can jump around two-dimensionally via the large amount of citrate anions as illustrated in Figure 15a. The TC molecules on the Ag colloidal nanoparticles that were kept in iced water and then adsorbed a small amount of citrate anions may be attracted one-dimensionally by the surface-plasmon-enhanced optical trapping potential well as illustrated in Figure 15b.

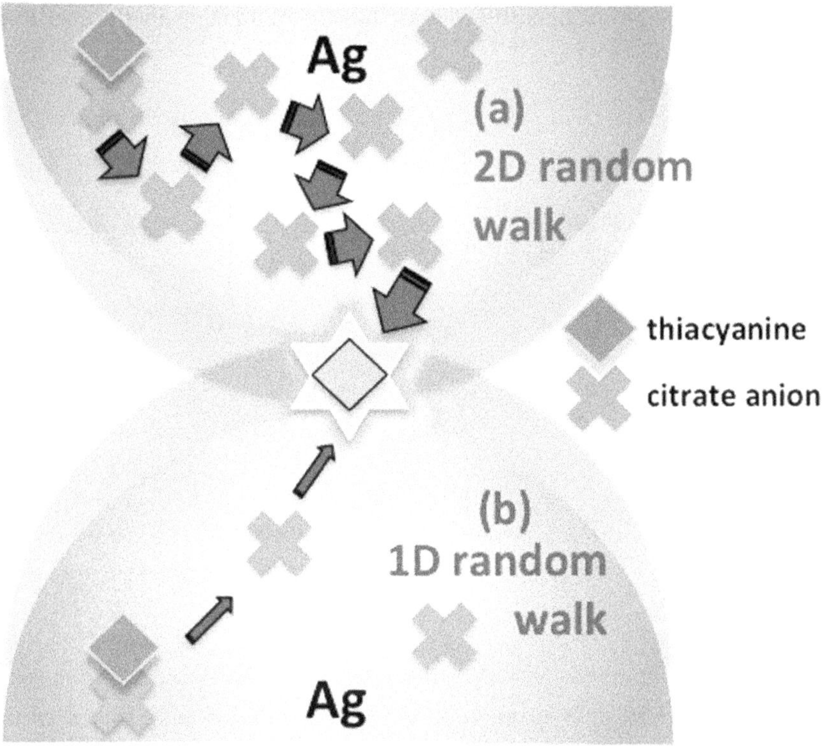

Figure 15. Schematic illustrations of molecular random walks on the Ag surface covered with (a) a large and (b) a small amount of citrate anions. (Reproduced with permission from ref. (24). Copyright 2015 Royal Chemical Society).

Truncation Time for the Dark Events

Physical Interpretation of the Truncation Times

The presence and absence of truncation for the dark and bright SERS events, respectively, are opposite to the blinking statistics in fluorescence from a single QD; the log-log plots for dark and bright fluorescence events display a line and a curve truncated at the tail, respectively (*64–68*). Figure 16 schematically illustrates the mechanism for the blinking of fluorescence from a single quantum dot and SERS. The fluorescence from a single QD originates from recombination of a photo-excited electron (e^-) in a conduction band and a hole (h^+) in a valence band as illustrated in Figure 16a. The fluorescence is quenched by the prevention of recombination through capturing the excited electron at a surface trap state (Figure 16b) (*63*, *68*). In contrast, SERS light is emitted from the molecule trapped at a junction of an Ag nanoaggregate where the EM field is greatly enhanced (Figure 16d) (*1–7*), and no SERS emerges from the molecule on the rest of the surface of the Ag nanoaggregate (Figure 16c). Thus, the mechanism for bright SERS events is similar to that of dark fluorescence events in a single QD, and the probability distributions of the blinking SERS contrast those of the blinking fluorescence from a single QD (*20, 21*).

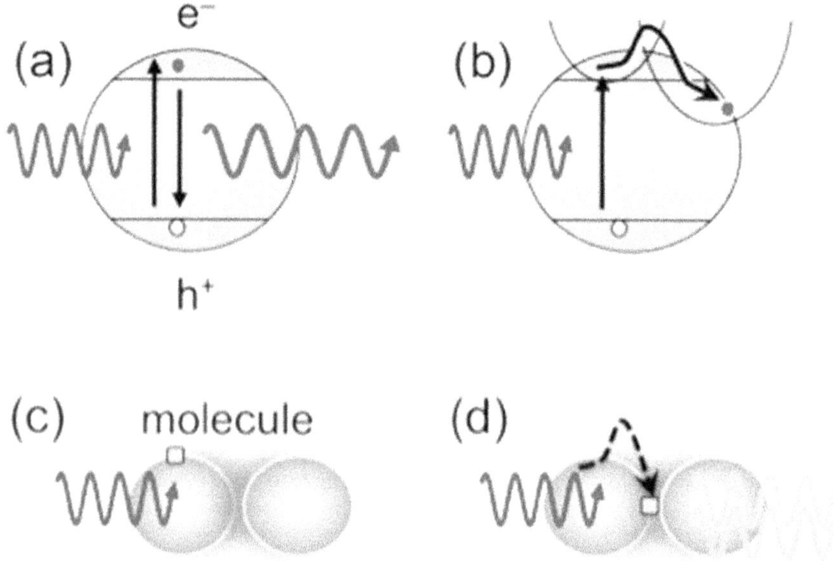

Figure 16. Schematic illustrations of the (a) bright and (b) dark fluorescence events for a single quantum dot, and those of (c) the dark and (d) bright SERS events for a single molecule adsorbed on an Ag nanoaggregate. (Reproduced with permission from ref. (20). Copyright 2010 Royal Chemical Society 20).

The truncation at the tail of the power law for bright or dark events is reproduced by a random walk on parabolic potential surfaces against the reaction coordinate for emissive and non-emissive states (*68*). In the quenching and re-emission of fluorescence from a single QD, the electron-hole pair is separated and reformed via an energy barrier between the photo-ionized and -excited states (Figure 16b), respectively. According to the differential equation for a classical random walk on a harmonic potential of a random walk-controlled electron-transfer model for blinking fluorescence from a single QD, the truncation time is given as

$$\tau = \frac{\Gamma}{\mathrm{E}_a/2k_BT} \qquad (6)$$

where E_a is an energy barrier between the emissive and non-emissive states, and Γ is the random walk time to overcome the energy barrier (*68*). When there is no energy barrier or a very long random walk time (slow random walk), the truncation time becomes much longer, namely, truncation does not occur. For dark fluorescence events from a single QD, the excited electron at the surface trap state moves more slowly than that in a conduction band, and thus no truncation occurs (*68*). Additionally, in bright SERS events, the random walk of the adsorbate molecule at the junction may be slower than that on the rest of the surface, and thus no truncation takes place (*20, 21*). At the SERS-active junction, the enhanced EM field due to LSPR can be used not only for single molecule spectroscopy, but also for single molecule optical trapping by the surface-plasmon-enhanced optical potential well around the junction (*79*).

To confirm the possibility of single molecule optical trapping, the potential around a gap in an Ag nanodimer was calculated (*22*). An optical trapping potential is given by $\alpha E^2/2$, where α is the polarizability of a molecule, and E is an electric field of light. The potential of 0.01 k_BT at 300 K is derived from the assumed parameters as follows: a polarizability of a single dye molecule (10^{-37} J^{-1} C^2 m^2) (*81*), an adequate SERS enhancement factor for single molecule detection ($E^4/E_0^4 = 10^{12}$, where E is a surface-plasmon-enhanced EM field, and E_0 is an original EM field of the excitation light), and the excitation laser intensity (200 W/cm^2, • E_0^2). The optical potential is much smaller than the thermal energy, although it can work as the trapping potential well to a single molecule not in a solution, but electrically adsorbed on a colloidal metal surface. Moreover, its effective polarizability on the metal is increased by the effect of a mirror image dipole (*82*). Indeed, it has been suggested that the potential of 0.1 k_BT works to immobilize an adsorbate molecule onto the junction by suppression of the blinking SERS at 77 K (*25*).

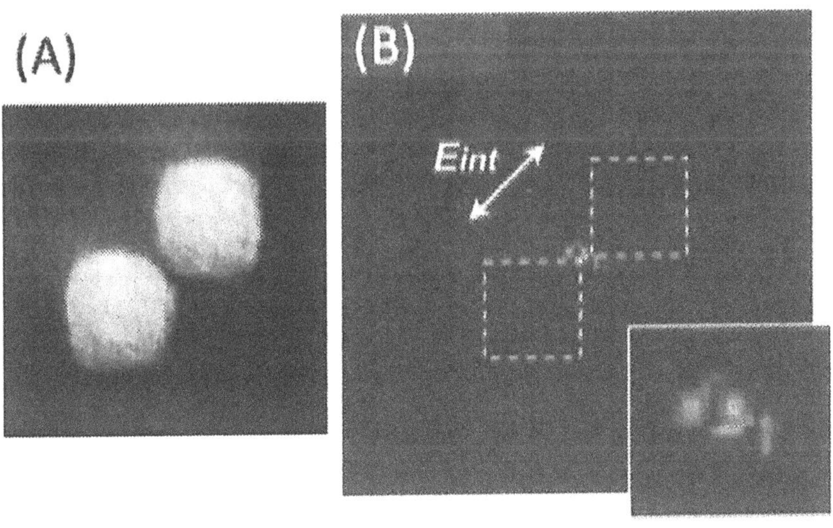

Figure 17. (A) SEM and (B) scattered near-field images of a pair of Au blocks (the length of the side is 100 nm) illuminated with Ti:sapphire laser light at 800 nm under a focused internal reflection. The inset in (B) is a magnified image around the gap of the dimer. (Reproduced with permission from ref. (85). Copyright 2012 Yamada Science Foundation).

For blinking SERS, a candidate for the energy barrier between emissive and non-emissive states (the junction and the rest of surface, respectively) may be the periodic distribution of the local EM field around the junction (*20*, *21*, *23*). For noble metal nanoparticles, a periodic pattern of the EM field has been calculated by a transfer matrix technique, Green function, and FDTD method (*80*, *83*, *84*). Figure 17 shows the periodic distribution of the EM field around the gap of Au nanoblocks that has recently been detected by a scattering-type scanning near-field microscope (*85*). It is thought that the periodic EM field originates from the coupling of multipolar surface plasmon resonance (*86*), which is induced by a high-energy excitation. Figure 18 depicts that the calculated EM fields around a gap of the Au disks show the periodic spatial distribution more clearly under the excitation at shorter wavelengths (*86*). Similar dependences on the periodic pattern and excitation wavelength appear in other FDTD calculations for a pair of Ag nanoparticles (*23*). It is noted that the dependence on the excitation wavelength is not correlated with the efficiency of LSPR. Figure 19 shows the truncation times, which may be shortened by the periodic EM field, at various LSPR wavelengths for single Ag nanoaggregates. They are almost constant despite the approach of the LSPR wavelength to the excitation wavelength (*23*).

Even in the periodic optical trapping potential well, the molecule in the meta-stable state (in the shallower bottoms) may not emit SERS light, but be optically trapped. It is because SERS is narrowly emitted from a single molecule even by the enormous enhancement at the hot spot and is quadratically weakened by a decrease in the optical trapping potential due to the two-fold enhancement (*74*, *75*, *87*). In short, the meta-stable state may be a non-emissive state. When the molecule quickly goes from the meta-stable state to the emissive state, it must overcome the energy-barrier between them. Thus, truncation in the power law for dark SERS events is induced by the periodic optical trapping potential well as the energy-barrier. In detail, τ is shortened by a large E_a (see Equation 6). On the other hand, Γ lengthens, and then τ becomes long, especially in the bright SERS (the very long τ means no truncation).

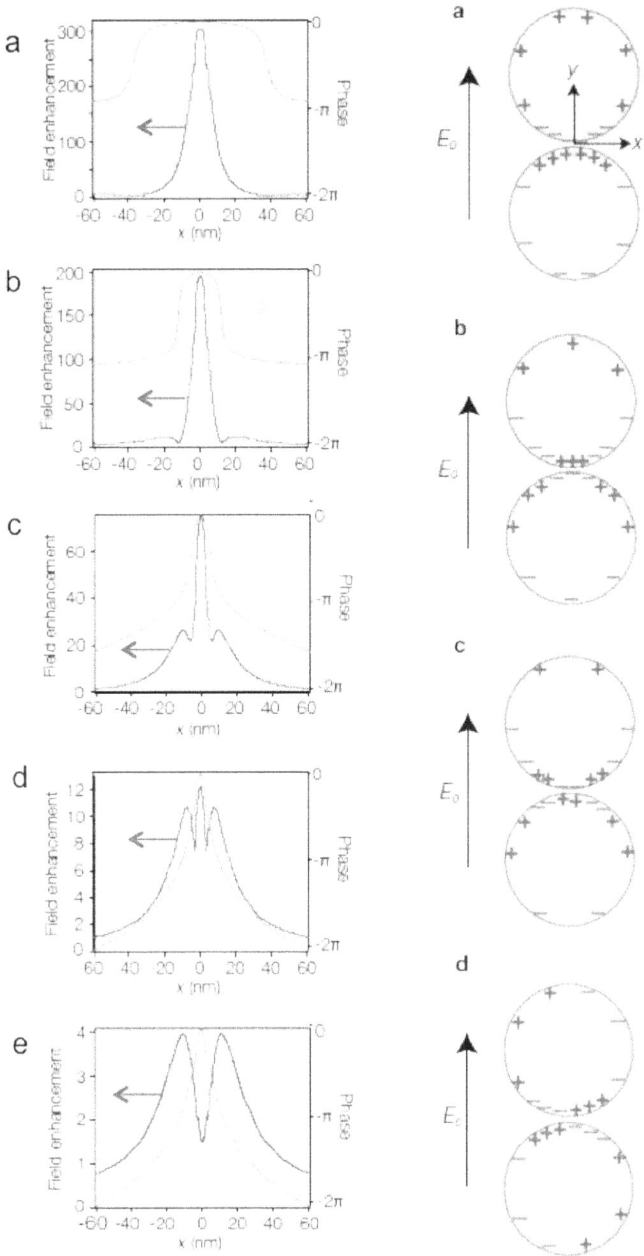

Figure 18. (Left) The cross-sectional profiles of the field amplitude (left y-axis) and phase (right y-axis) within the gap along the x-axis at an excitation energy of (a) 1.29, (b) 1.82, (c) 2.14, (d) 2.29, and (e) 2.38 eV. (Right) Charge distributions and field profiles of individual multipolar plasmonic modes. (Reproduced with permission from ref. (86). Copyright 2012 Nature Publishing Group).

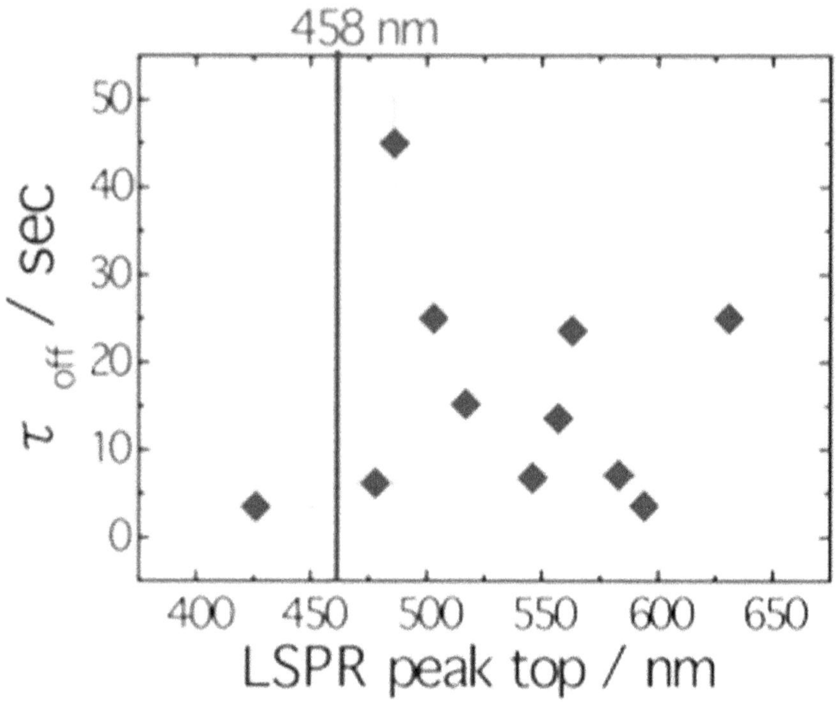

Figure 19. Truncation times of the truncated power law that reproduces the probability distributions against duration at various LSPR wavelengths for single Ag nanoaggregates with adsorbed anionic TC excited at 458 nm. (Reproduced with permission from ref. (23). Copyright 2013 American Chemical Society).

Effect of Periodic Enhanced Electromagnetic Field

Dependence on Excitation Light Intensity

Figure 20 shows the histograms of the truncation times in the power law that reproduces the probability distributions of dark SERS events against their durations under various laser intensities (11, 14, and 17 mW corresponds to power densities of 110, 140, and 170 W/cm^2, respectively) (*21*). The truncation times were almost constant despite changing the excitation laser intensity from 11 to 17 mW. It is noted that this differs from the truncation times for a single QD, which are shortened by a photo-induced random walk of the excited electron due to the higher excitation laser intensity (*64*, *66*).

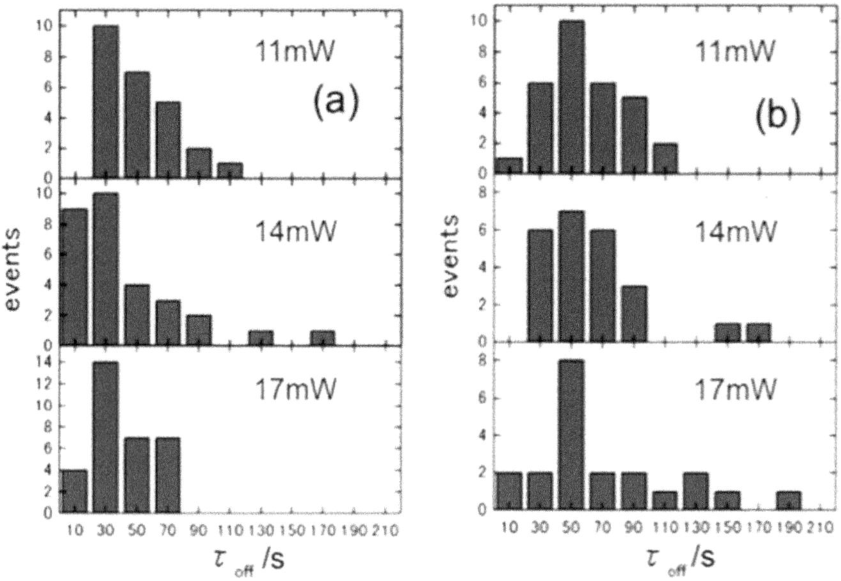

Figure 20. Histograms for the truncation times of the truncated power law that reproduces the probability distributions of (a) 514 or (b) 568-nm-excited dark SERS events against their durations for TCC adsorbed on single Ag nanoaggregates excited under the laser intensities of 11, 14, and 17 mW. (Reproduced with permission from ref. (21). Copyright 2011 Royal Chemical Society).

As discussed above, the truncation time is shortened by the high-energy barrier and a short random walk time to overwhelm the barrier. If the blinking SERS was affected by a photo-independent energy barrier, the truncation times were shortened by a higher excitation laser intensity through the photo-dependent random walk of the adsorbate like a single QD (*64*, *66*). However, the truncation times for SERS are then constant. Thus, the energy barrier may depend on excitation light such as the periodic surface-plasmon-enhanced EM field. The energy barrier due to the periodic EM field becomes higher as the laser intensity, the original EM field of excitation light, increases. It could allow truncation to occur at a shorter tail of the power law. However, the random walk of an adsorbate molecule is suppressed by a deeper optical trapping potential well due to a stronger surface-plasmon-enhanced EM field. A longer random walk time Γ (slower random walk) probably cancels out the higher-energy barrier E_a with a higher laser intensity as represented by Equation 6 ($\tau \bullet \Gamma/E_a$). Thus, the observed constant truncation times under various laser intensities can be explained by the energy barrier due to the periodic EM field.

The blinking SERS peaks of Fe-protoporphyrin IX (10^{-11} M) on Ag excited at 514 nm have already been analyzed by the power law without an exponential function (see Figure 10) (*19*). Thus, truncation in the power law could not be observed. The reason may be that time-resolved SERS spectra cannot have high time resolution for an accurate fitting by a truncated power law in a log-log plot rather than a semi-logarithm plot. Possibly, the strong laser power density (≤ 1 mW (*19*), which likely corresponds to a power density of ≤ 130 kW/cm^2) may change the meta-stable state into SERS-emissive state, and then the energy barrier, which induces truncation, between the emissive and non-emissive states will vanish.

Dependence on Excitation Wavelength

Figure 20a and 20b also represent that the maxima in the histograms of the truncation times were observed to be ~30 s at 514 nm and ~50 s at 568 nm, respectively (*23*). This is different to the dependence of blinking fluorescence on the excitation wavelength from a single QD; the truncation times for the bright fluorescence events were similar except for near-ultraviolet excitation (*65*). It was thought that photo-induced desorption might be accelerated by excitation at a shorter wavelength than an absorption band of a TCC dimer at 505 nm and that of the monomer at 550 nm (*88–90*), and the truncation time would be shortened then.

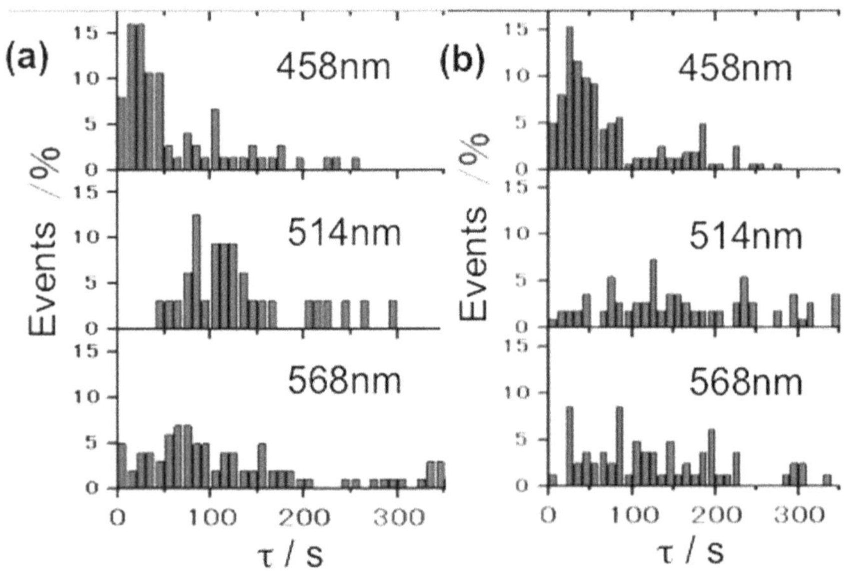

Figure 21. Histograms for the truncation times of the truncated power law that reproduces the probability distributions of dark SERS events against their durations for (a) 10 or (b) 25 μM zwitterionic TC adsorbed on single Ag nanoaggregates excited at 458, 514, and 568 nm (23).

Unlike a TCC molecule, a TC molecule barely absorbs the excitation light whose wavelength is longer than 458 nm (*39*). Thus, the dependence of truncation on the excitation wavelength for TC is not attributed to the photo-induced desorption. Figure 21 shows the histograms of the truncation times in the power law for the dark SERS events from the TC adsorbed on 32–163 single Ag nanoaggregates excited at 458, 514, and 568 nm (*23*). Despite the different concentrations of TC (Figures 21a and 21b correspond to 10 and 25 μM TC, respectively), the maxima of the histograms by the excitation at 514 and 568 nm appear around 100–150 s, while the maxima at 458 nm appear around 0–50 s. The averages and medians of the truncation times were increased by excitation at long wavelengths of 514 and 568 nm compared to 458 nm, as summarized in Table 2. Moreover, the probability distribution for the dark SERS event excited at the long excitation wavelengths became difficult to be reproduced using the truncated power law; namely, the truncation times were often deduced to be very long with large errors. By the excitations at 568 and 514 nm, the percentages of events whose probability distribution cannot be reproduced using the truncated power law are larger than those at 458 nm as summarized in Table 2. Consequently, the truncation time is shortened by the excitation at short wavelengths via the periodic optical trapping potential well, which originates from the coupling of the multipolar surface plasmon resonance induced by high-energy excitation (*23*, *86*).

Table 2. Averages and medians of the truncation times, and failure percentages in reproducing the probability distribution for dark SERS events by a truncated power law at various excitation wavelengths, at different concentrations of zwitterionic TC (*23*); from the Ag nanoaggregates kept at room temperature and in iced water, which were covered with large and small amounts of citrate anions, respectively (*24*), and at different concentrations of NaCl

$\tau_{av.\ (R.T.;\ 100mM)}$	458 nm	514 nm	568 nm
10 μM	74 ± 10 s	162 ± 26 s	186 ± 19 s
25 μM	85 ± 7 s	256 ± 20 s	221 ± 24 s

$\tau_{av.\ (458nm;\ 25\mu M)}$	R. T.	Iced water
100 mM	64 ± 5 s	63 ± 6 s
10 mM	98 ± 10 s	

$\tau_{med.\ (R.T.;\ 100mM)}$	458 nm	514 nm	568 nm
10 μM	40 s	119 s	117 s
25 μM	51 s	190 s	142 s

$\tau_{med.(458nm;\ 25\mu M)}$	R. T.	Iced water
100 mM	36 s	42 s
10 mM	62 s	

The fail $_{(R.T.;\ 100mM)}$	458 nm	514 nm	568 nm
10 μM	7% (=6/81)	29 % (=13/45)	34 % (=53/155)
25 μM	8% (=15/178)	29 % (=44/154)	17 % (=17/99)

The fail $_{(458nm;\ 25\mu M)}$	R. T.	Iced water
100 mM	9 % (=19/213)	29 % (=39/135)
10 mM	22 % (=39/180)	

Effect of the Adsorption on the Ag Surface

Dependence on the NaCl Concentration

Figure 22 shows the histograms of the truncation times derived from 139 and 196 single Ag nanoaggregates with NaCl at 10 and 100 mM, respectively. The maxima in the histograms were located at 30–40 s and around 10 s from the Ag nanoaggregate with 10 and 100 mM NaCl, respectively. Similarly, the average and median of the truncation times were observed to be 98 and 62 s at 10 mM, or 64 and 36 s at 100 mM, respectively. At 10 mM, moreover, the percentages of the events whose probability distributions cannot be reproduced using the truncated power law are larger than those at 100 nm. Thus, the truncation times at 100 mM (NaCl) are shorter than those at 10 mM.

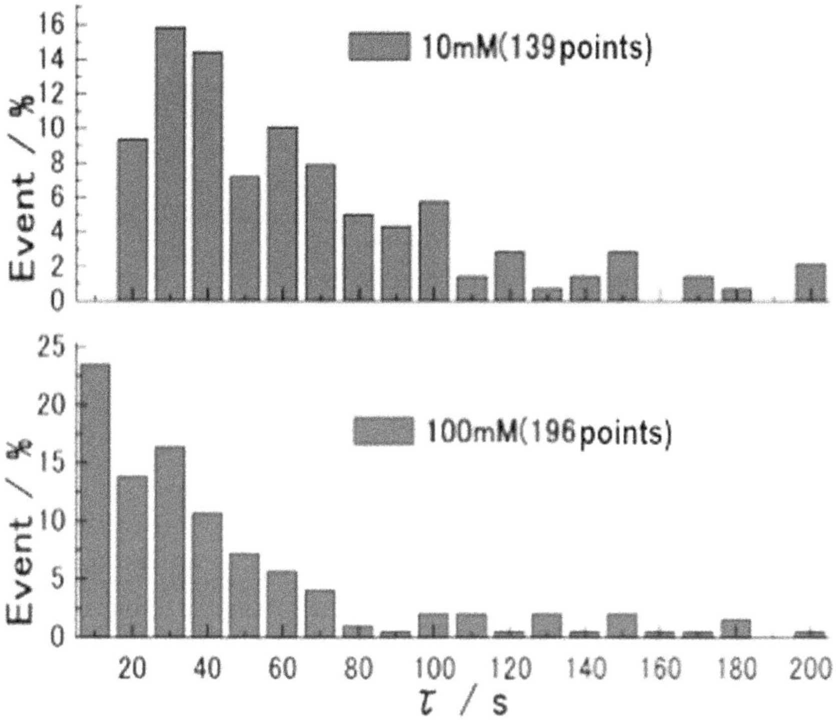

Figure 22. Histograms for the truncation times of the truncated power law that reproduces the probability distributions of dark SERS events against their durations for zwitterionic TC adsorbed on single Ag nanoaggregates with 100 and 10 mM NaCl.

The absorption spectra of the supernatant of the Ag colloidal suspension mixed with a TC aqueous solution shows that the adsorbates on the Ag surface at 10 mM (NaCl) are a similar value to those at 100 mM (data not shown). Therefore, the dependence of τ on NaCl concentration is not induced through the dye concentration. At 100 mM (NaCl), Ag nanoparticles may grow larger, and then the LSPR band likely appears at longer wavelengths. However, τ is independent from the LSPR. Thus, the reason for the dependence of the truncation time on the NaCl concentration may be that the number of junctions per Ag nanoaggregate is increased by the coherence of the nanoparticles; for example, a dimer has one junction, a trimer can have three junctions, and a tetramer can be arranged as a tetrahedron and can have six junctions. On the larger Ag nanoaggregate with 100 mM NaCl, the TC molecule may reach the junction more quickly.

Dependence on the Citrate Anion Concentration

Figure 23 shows the histograms of the truncation times derived from 96 and 194 single Ag nanoaggregates that were kept in iced water at room temperature overnight, which are covered with smaller and larger amounts of citrate anions, respectively (*24*). The maximum of the histogram of the truncation times from the Ag nanoaggregates covered with the larger amount of citrate anions appears at a shorter time than that covered with a smaller amount of them. Additionally, the medians of the truncation times show this trend as summarized in Table 2. The averages of the truncation times are similar within their standard errors as expressed in Table 2. However, the probability distributions of dark SERS events from the Ag nanoaggregates covered with a smaller amount of citrate anions cannot be reproduced more often than those covered with a larger amount of citrate anions by the truncated power law (Table 2); namely, the percentage of very long truncation times, which were not used for the derivation of the average, of the Ag nanoaggregates covered with a smaller amount of citrate anions are higher than those covered with a larger amount of them. Thus, it is concluded that the truncation times of the Ag nanoaggregates covered with a larger amount of citrate anions are shorter than those covered with a smaller amount of them. This reason is considered in the next section.

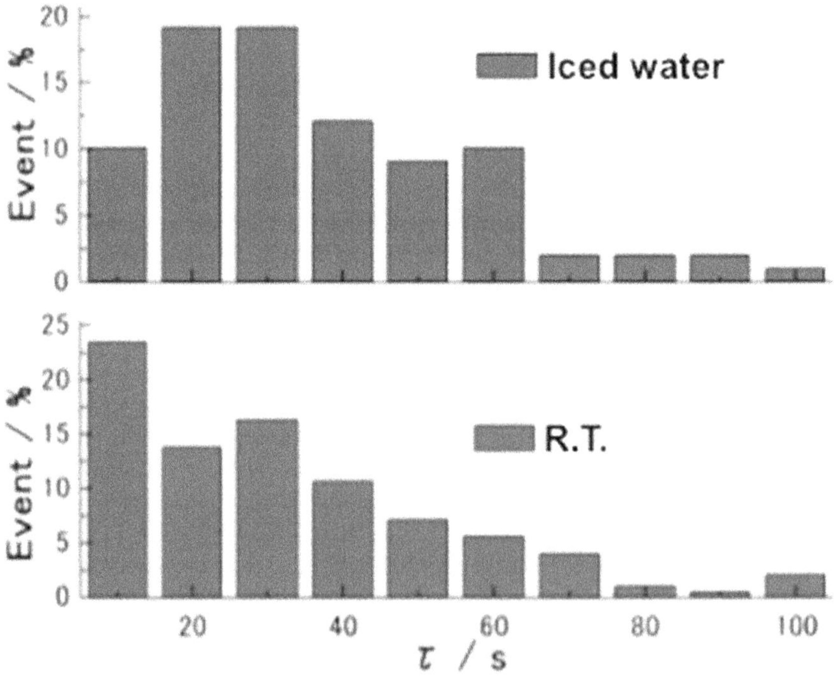

Figure 23. Histograms for the truncation times of the truncated power law that reproduces the probability distributions of dark SERS events against their durations for zwitterionic TC adsorbed on single Ag nanoaggregates kept at room temperature and in iced water, which are covered with large and small amounts of citrate anions (24).

Dependence on the Thiacyanine Concentration

An aqueous solution of the twitterionic TC (10 or 25 μM), a NaCl aqueous solution (100 mM), and an Ag colloidal suspension were mixed in a volume ratio of 1:1:2 (*23*). Figure 24 shows the histograms of the truncation times derived from 32–163 single Ag nanoaggregates with 10 and 25 μM TC. Figure 24a indicates a larger percentage of the events of τ = 50–100 s by the excitation at 458 nm for 25 μM than that for 10 μM. Figures 24b and 24c reveal that the maxima of the histograms by the excitation at 514 and 568 nm for 25 μM appear at a longer truncation time than those for 10 μM, respectively. Indeed, the averages and medians of the truncation times were increased at the higher concentrations from 10 to 25 μM as summarized in Table 2. Thus, the truncation times were lengthened at a higher concentration of TC (*23*).

This is consistent with the dependence on citrate anion concentration; the truncation times from the Ag nanoaggregates that had been kept at room temperature were shorter than those in iced water due to the increase of the citrate anion on the Ag surface (*24*). Therefore, the truncation times are shortened by an increase in the ratio of the stabilizer to the adsorbate. The reason may be that the adsorbate molecule on the Ag surface can reach the junction more quickly via the high amount of stabilizers that act as hopping sites.

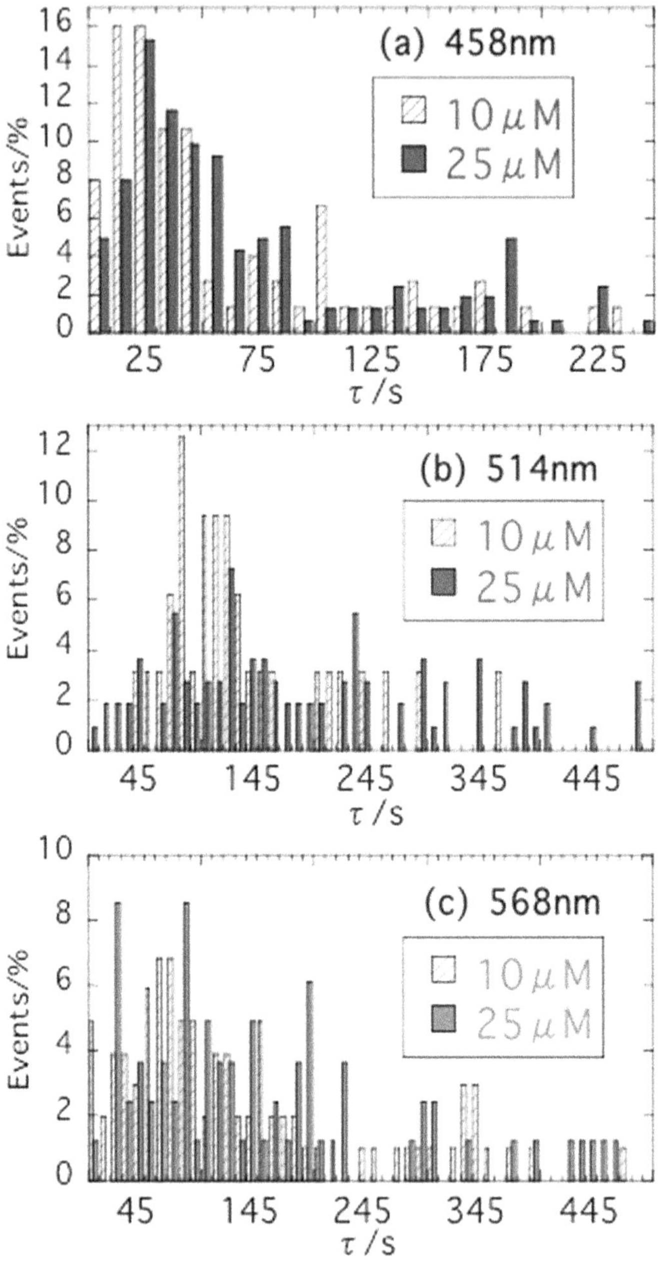

Figure 24. Histograms for the truncation times of the truncated power law that reproduces the probability distributions of dark SERS events against their durations for 10 and 25 μM zwitterionic TC adsorbed on single Ag nanoaggregates excited at (a) 458, (b) 514, and (c) 568 nm. (Reproduced with permission from ref. (23). Copyright 2013 American Chemical Society).

Difference between Cationic and Twitterionic Thiacyanine

For cationic TC, the truncation times range from 5 to 25 s except for a few cases with excitations at 458 nm as shown in Figure 19 (*23*) and Figure 4 in the ref. (*20*), while the average and median truncation times for the zwitterionic TC excited at 458 nm were 64–98 and 36–62 s, respectively, as summarized in Table 2. In the former case, an aqueous solution of the cationic TC (2 and 4 μM), a NaCl aqueous solution (1 mM), and an Ag colloidal suspension were mixed in a volume ratio of 1 : 1 : 2 (*20*, *23*). In the latter case, an aqueous solution of the zwitterionic TC (10 and 25 μM), a NaCl aqueous solution (10 and 100 mM), and an Ag colloidal suspension were mixed in a volume ratio of 1 : 1 : 2 (*23*, *24*).

As discussed above, the dependence on concentration indicates that the truncation time is lengthened at higher and lower concentrations of adsorbate and NaCl, respectively. The effects on the truncation time at a higher concentration of zwitterionic TC and NaCl probably cancel each other out. Furthermore, the zwitterionic TC molecules are barely adsorbed (0.46 μM) on the Ag nanoaggregates by the addition of the solution (25 μM) in a volume ratio of 1/4 (*24*). Thus, the different truncation times may be attributed to an electric charge of the TCs. It seems that the zwitterionic TC, which has one $>N^{+}=$ and two $-SO_3^{-}$, is repulsive and then could move quickly on the Ag surface covered with the citrate anions. Actually, the zwitterionic TC may hardly move via the citrate anions on the Ag surface because it is repelled by the surrounding anions. In contrast, the cationic TC can move on the Ag quickly because it is electrically attracted by the citrate anion. Thus, electric charges of the adsorbed molecules likely influence the truncation times through the random walk times on the citrate-reduced Ag surface.

Summary

In the blinking SERS, the power law exponents in the bright and dark SERS are simultaneously increased and decreased, respectively, by effective LSPR through the excitation at the wavelength closer to the LSPR bands (*22*), excitation at the longer wavelength (568 nm) (*23*), under the more intense excitation light (*22*), and/or at the lower concentration of NaCl (10 mM). The opposite trends may originate from the 'hot site', where SERS is emitted, become deeper and wider, while the rest of the Ag surface, where no SERS is emitted, becomes smaller through more effective LSPR. Then, the power law exponents approach −1.5 under the effective LSPR condition (*22*, *23*). This value is derived from the one-dimensional random walk of the molecule due to the restraint through the surface-plasmon-enhanced optical trapping force around the junction. This is similar to those in the bright and dark fluorescence from a single QD (*63–66*); the emissive photo-excited electron and hole pair is restrained by Coulomb attraction. Conversely, only the α_{off} are changed by a different amount of the citrate anions, which cover the colloidal Ag nanoaggregates, while the α_{on} display a similar distribution because the Ag nanoaggregates have the same size and show the same LSPR band (*24*). On the Ag nanoaggregates with a large amount of the citrate anion, the α_{off} approached –1. This value may be derived from the two-dimensional random walk on the metallic surface via many hopping site, namely, a large amount of the citrate anions.

The truncation at the tail of the power law originates from overcoming the energy-barrier between the emissive and non-emissive states (*68*). In the case of the blinking SERS, the energy-barrier may be attributed to the periodic optical trapping potential well. It is noted that the truncation times are independent from the efficiency in LSPR; they are almost constant despite the approach of the LSPR wavelength to the excitation wavelength or the higher excitation laser intensity (*21*, *23*). The reason may be explained as follows. The molecular random walk will likely become slower by the deeper surface-plasmon-enhanced optical trapping potential well. The longer random walk time Γ probably cancels out the higher energy barrier E_a through the effective LSPR, and then the truncation times (τ • Γ/E_a) are almost constant. Additionally, the truncation times are shortened by excitation at shorter wavelengths (*23*). This result suggests that the energy-barrier is attributed to the periodic optical trapping potential well due to the coupling of the multipolar surface plasmon resonance, which is induced by the excitation at short wavelengths (*85*, *86*). The dependence of the truncation times on the NaCl concentration, which are shortened at high concentrations, cannot be explained by the efficiency in LSPR unlike that of the power law exponents. The reason may be that the number of junctions can be increased in the larger nanoaggregate by the addition of NaCl. By increasing the ratio of the citrate anion to the ionic dye on the Ag surface, the truncation times were shortened (*23*, *24*). The reason may be that the molecule can reach the junction more quickly via larger number of the citrate anion. For cationic TC, the truncation times may be shorter than those for zwitterionic TC (*20*, *23*, *24*). It indicates that the citrate anions on the Ag surface work as hopping sites for the cationic TC and as obstacles for the zwitterionic TC. It is noteworthy that the description is contrary to the idea that the zwitterionic TC is more repulsive and then could move quickly on the Ag surface covered with the citrate anions.

Consequently, the excitation light intensity and wavelength affect the dimension of the molecular random walk on the Ag surface via the surface-plasmon-enhanced optical trapping potential well. The dependence of the truncation time on the excitation light intensity and wavelength suggests the periodicity of the enhanced EM field. The concentration of NaCl influences the dimension of the molecular random walk via the efficiency of LSPR for Ag nanoaggregates, while the dependence of the truncation time can be explained not by the enhanced EM field due to LSPR, but by the condition of the Ag surface via the speed of the molecular random walk. On the other hand, the concentration of citrate anions affects the dimension and speed of the molecular random walk in terms of the adsorption of an ionic dye molecule on the Ag surface. By the dependence of the truncation time on the concentration of NaCl, citrate anions, and dye molecules, it can be estimated how fast the molecule moves on the Ag surface, except for at the junction. From SERS and luminescence on a single Ag nanoaggregate, the centroid positions have been revealed using super-resolution imaging (*46*, *47*), although, it may be difficult to detect the molecular behavior on the Ag surface except for the junction. Thus, the dependence of blinking SERS on various conditions can be divided into those of the power law exponents and the truncation times by the truncated power law analysis. Then, the complex behavior may be separated into simple factors, the dimension and speed of the molecular random walk through the enhanced EM field, and the adsorption on the metal. It is useful to explore the essence of blinking SERS.

Acknowledgments

I would like to express appreciation Prof. Y. Ozaki of Kwansei Gakuin University for fruitful discussion and his encouragement. The author is grateful to Dr. T. Itoh of National Institute of Advanced Industrial Science and Technology (AIST) for productive discussion. I also thank Mr. Y. Tanaka, Ms. A. Enogaki, Mr. T. Nagahiro, and Mr. D. Araki, who were students of Kwansei Gakuin University, for their contributions to these studies. This work was supported by KAKENHI (Grant-in-Aid for Scientific Research C) (No. 16K05671).

References

1. Aroca, R. *Surface-Enhanced Vibrational Spectroscopy*; John Wiley & Sons Ltd.: Chichester, U.K., 2006.
2. Kneipp, K.; Moskovits, M.; Kneipp, H. *Surface-Enhanced Raman Scattering: Physics and Applications*; Springer: Berlin, 2006.
3. Le Ru, E. C.; Etchegoin P. G. *Principles of Surface-Enhanced Raman Spectroscopy and Related Plasmonic Effects*; Elsevier: Amsterdam, The Netherlands, 2009.
4. Itoh, T.; Sujith, A.; Ozaki, Y. In *Frontiers of Molecular Spectroscopy*; Laane, J., Ed.; Elsevier: Amsterdam, The Netherlands, 2009; pp289−320.
5. Kitahama, Y.; Hossain, M. K.; Itoh, T.; Sujith, A.; Han, X. X.; Ozaki, Y. In *Raman, Infrared, and Near-Infrared Chemical Imaging*; Šašić, S., Ozaki, Y., Eds.; Wiley: Hoboken, NJ, 2010; pp 285−299.
6. Kitahama, Y.; Itoh, T.; Pienpinijtham, P.; Ekgasit, S.; Han, X. X.; Ozaki, Y. In *Functional Nanoparticles for Bioanalysis, Nanomedicine and Bioelectronic Devices Vol. 2*; Hepel, M., Zhong, C. J., Eds.; ACS Symposium Series 1113; American Chemical Society: Washington, DC, 2013; pp 181−234.
7. Kitahama, Y.; Ozaki, Y. In *Frontiers of Surface-Enhanced Raman Scattering: Single-Nanoparticles and Single Cells*; Ozaki, Y., Kneipp, K., Aroca, R., Eds.; Wiley: Chichester, 2014; pp 107−137.
8. Qian, X. M.; Nie, S. M. *Chem. Soc. Rev.* **2008**, *37*, 912–920.
9. Pieczonka, N. P. W.; Aroca, R. F. *Chem. Soc. Rev.* **2008**, *37*, 946–954.
10. Kneipp, J.; Kneipp, H.; Kneipp, K. *Chem. Soc. Rev.* **2008**, *37*, 1052–1060.
11. Etchegoin, P. G.; Le Ru, E. C. *Phys. Chem. Chem. Phys.* **2008**, *10*, 6079–6089.
12. Itoh, T.; Yamamoto, Y. S. *Analyst* **2016**, *141*, 5000–5009.
13. Kneipp, K.; Wang, Y.; Kneipp, H.; Perelman, L. T.; Itzkan, I.; Dasari, R. R.; Feld, M. S. *Phys. Rev. Lett.* **1997**, *78*, 1667–1670.
14. Nie, S.; Emory, S. R. *Science* **1997**, *275*, 1102–1106.

15. Habuchi, S.; Cotlet, M.; Gronheid, R.; Dirix, G.; Michiels, J.; Vanderleyden, J.; De Schryver, F. C.; Hofkens, J. *J. Am. Chem. Soc.* **2003**, *125*, 8446–8447.
16. Weiss, A.; Haran, G. *J. Phys. Chem. B* **2001**, *105*, 12348–12354.
17. Emory, S. R.; Jensen, R. A.; Wenda, T.; Han, M.; Nie, S. *Faraday Discuss.* **2006**, *132*, 249–259.
18. Shegai, T. O.; Haran, G. *J. Phys. Chem. B* **2006**, *110*, 2459–2461.
19. Bizzarri, A. R.; Cannistraro, S. *Phys. Rev. Lett.* **2005**, *94*, 068303-1–068303-4.
20. Kitahama, Y.; Tanaka, Y.; Itoh, T.; Ozaki, Y. *Phys. Chem. Chem. Phys.* **2010**, *12*, 7457–7460.
21. Kitahama, Y.; Tanaka, Y.; Itoh, T.; Ozaki, Y. *Chem. Commun.* **2011**, *47*, 3888–3890.
22. Kitahama, Y.; Tanaka, Y.; Itoh, T.; Ozaki, Y. *Phys. Chem. Chem. Phys.* **2011**, *13*, 7439–7448.
23. Kitahama, Y.; Enogaki, A.; Tanaka, Y.; Itoh, T.; Ozaki, Y. *J. Phys. Chem. C* **2013**, *117*, 9397–9403.
24. Kitahama, Y.; Araki, D.; Yamamoto, Y. S.; Itoh, T.; Ozaki, Y. *Phys. Chem. Chem. Phys.* **2015**, *17*, 21204–21210.
25. Maruyama, Y.; Ishikawa, M.; Futamata, M. *J. Phys. Chem. B* **2004**, *108*, 673–678.
26. Lombardi, J. R.; Birke, R. L.; Haran, G. *J. Phys. Chem. C* **2011**, *115*, 4540–4545.
27. Itoh, T.; Hashimoto, K.; Biju, V.; Ishikawa, M.; Wood, B. R.; Ozaki, Y. *J. Phys. Chem. B* **2006**, *110*, 9579–9585.
28. Bjerneld, E. J.; Földes-Papp, Z.; Käll, M.; Rigler, R. *J. Phys. Chem. B* **2002**, *106*, 1213–1218.
29. Bizzarri, A. R.; Cannistraro, S. *Appl. Spectrosc.* **2002**, *56*, 1531–1537.
30. Etchegoin, P.; Liem, H.; Maher, R. C.; Cohen, L. F.; Brown, R. J. C.; Milton, M. J. T.; Gallop, J. C. *Chem. Phys. Lett.* **2003**, *367*, 223–229.
31. Delfino, I.; Bizzarri, A. R.; Cannistraro, S. *Biophys. Chem.* **2005**, *113*, 41–51.
32. Šašić, S.; Itoh, T.; Ozaki, Y. *J. Raman Spectrosc.* **2005**, *36*, 593–599.
33. Ruan, C.; Wang, W.; Gu, B. *J. Raman Spectrosc.* **2007**, *38*, 568–573.
34. Doering, W. E.; Nie, S. *J. Phys. Chem. B* **2002**, *106*, 311–317.
35. Bizzarri, A. R.; Cannistraro, S. *J. Phys. Chem. B* **2005**, *109*, 16571–16574.
36. Bizzarri, A. R.; Cannistraro, S. *Phys. Chem. Chem. Phys.* **2007**, *9*, 5315–5319.
37. Le Ru, E. C.; Etchegoin, P. G.; Meyer, M. *J. Chem. Phys.* **2006**, *125*, 204701-1–204701-13.
38. Etchegoin, P. G.; Lacharmoise, P. D.; Le Ru, E. C. *Anal. Chem.* **2009**, *81*, 682–688.
39. Kitahama, Y.; Tanaka, Y.; Itoh, T.; Ishikawa, M.; Ozaki, Y. *Bull. Chem. Soc. Jpn.* **2009**, *82*, 1126–1132.
40. Wang, Z.; Rothberg, L. J. *J. Phys. Chem. B* **2005**, *109*, 3387–3391.
41. Michaels, A. M.; Nirmal, M.; Brus, L. E. *J. Am. Chem. Soc.* **1999**, *121*, 9932–9939.

42. Itoh, T.; Iga, M.; Tamaru, H.; Yoshida, K.; Biju, V.; Ishikawa, M. *J. Chem. Phys.* **2012**, *136*, 024703-1–024703-7.
43. Peyser-Capadona, L.; Zheng, J.; González, J. I.; Lee, T.-H.; Patel, S. A.; Dickson, R. M. *Phys. Rev. Lett.* **2005**, *94*, 058301-1–058301-4.
44. Michaels, A. M.; Jiang, J.; Brus, L. E. *J. Phys. Chem. B* **2000**, *104*, 11965–11971.
45. Bosnick, K. A.; Jiang, J.; Brus, L. E. *J. Phys. Chem. B* **2002**, *106*, 8096–8099.
46. Stranahan, S. M.; Willets, K. A. *Nano Lett.* **2010**, *10*, 3777–3784.
47. Weber, M. L.; Litz, J. P.; Masiello, D. J.; Willets, K. A. *ACS Nano* **2012**, *6*, 1839–1848.
48. Le Ru, E. C.; Meyer, M.; Etchegoin, P. G. *J. Phys. Chem. B* **2006**, *110*, 1944–1948.
49. Dieringer, J. A.; Lettan, R. B.; Scheidt, K. A.; Van Duyne, R. P. *J. Am. Chem. Soc.* **2007**, *129*, 16249–16256.
50. Goulet, P. J. G.; Pieczonka, N. P. W.; Aroca, R. F. *J. Raman Spectrosc.* **2005**, *36*, 574–580.
51. Sawai, Y.; Takimoto, B.; Nabika, H.; Ajito, K.; Murakoshi, K. *J. Am. Chem. Soc.* **2007**, *129*, 1658–1662.
52. Takase, M.; Sawai, Y.; Nabika, H.; Murakoshi, K. *J. Photochem. Photobiol. A* **2011**, *221*, 169–174.
53. Nagasawa, F.; Takase, M.; Nabika, H.; Murakoshi, K. *Chem. Commun.* **2011**, *47*, 4514–4516.
54. Kudelski, A.; Pettinger, B. *Chem. Phys. Lett.* **2000**, *321*, 356–362.
55. Kudelski, A. *Vib. Spectrosc.* **2006**, *41*, 83–89.
56. Domke, K. F.; Zhang, D.; Pettinger, B. *J. Phys. Chem. C* **2007**, *111*, 8611–8616.
57. Itoh, K.; Kudryashov, I.; Yamagata, J.; Nishizawa, T.; Fujii, M.; Osaka, N. *J. Phys. Chem. B* **2005**, *109*, 271–276.
58. Szeghalmi, A.; Kaminskyj, S.; Rösch, P.; Popp, J.; Gough, K. M. *J. Phys. Chem. B* **2007**, *111*, 12916–12924.
59. Chaigneau, M.; Picardi, G.; Ossikovski, R. *Surf. Sci.* **2010**, *604*, 701–705.
60. Itoh, T.; Yamamoto, Y. S.; Biju, V.; Tamaru, H.; Wakida, S. *AIP Adv.* **2015**, *5*, 127113-1–127113-10.
61. Vantasin, S.; Ji, W.; Tanaka, Y.; Kitahama, Y.; Wang, M.; Wongravee, K.; Gatemala, H.; Ekgasit, S.; Ozaki, Y. *Angew. Chem. Int. Ed.* **2016**, *55*, 8391–8395.
62. Fujita, K.; Ishitobi, S.; Hamada, K.; Smith, N. I.; Taguchi, A.; Inouye, Y.; Kawata, S. *J. Biomed. Opt.* **2009**, *14*, 024038-1–024038-7.
63. Cichos, F.; von Borczyskowski, C.; Orrit, M. *Curr. Opin. Colloid Interface Sci.* **2007**, *12*, 272–284.
64. Shimizu, K. T.; Neuhauser, R. G.; Leatherdale, C. A.; Empedocles, S. A.; Woo, W. K.; Bawendi, M. G. *Phys. Rev. B* **2001**, *63*, 205316-1–205316-5.
65. Crouch, C. H.; Mohr, R.; Emmons, T.; Wang, S.; Drndic, M. *J. Phys. Chem. C* **2009**, *113*, 12059–12066.
66. Peterson, J. J.; Nesbitt, D. J. *Nano Lett.* **2009**, *9*, 338–392.

67. Crouch, C. H.; Sauter, O.; Wu, X.; Purcell, R.; Querner, C.; Drndic, M.; Pelton, M. *Nano Lett.* **2010**, *10*, 1692–1698.
68. Tang, J.; Marcus, R. A. *J. Chem. Phys.* **2005**, *123*, 054704-1–054704-12.
69. Krichevsky, O.; Bonnet, G. *Rep. Prog. Phys.* **2002**, *65*, 251–297.
70. Hess, S. T.; Huang, S.; Heikal, A. A.; Webb, W. W. *Biochemistry* **2002**, *41*, 697–705.
71. Schroeder, M. *Fractals, Chaos, Power Laws*; W. H. Freeman & Co.: New York, 1991.
72. Lee, P. C.; Meisel, D. J. *J. Phys. Chem.* **1982**, *86*, 3391–3395.
73. Galloway, C. M.; Etchegoin, P. G.; Le Ru, E. C. *Phys. Rev. Lett.* **2009**, *103*, 063003-1–063003-4.
74. Yoshida, K.; Itoh, T.; Tamaru, H.; Biju, V.; Ishikawa, M.; Ozaki, Y. *Phys. Rev. B* **2010**, *81*, 115406-1–115406-9.
75. Itoh, T.; Yoshida, K.; Biju, V.; Kikkawa, Y.; Ishikawa, M.; Ozaki, Y. *Phys. Rev. B* **2007**, *76*, 085405-1–085405-5.
76. Itoh, T.; Biju, V.; Ishikawa, M.; Kikkawa, Y.; Hashimoto, K.; Ikehata, A.; Ozaki, Y. *J. Chem. Phys.* **2006**, *124*, 134708-1–134708-6.
77. Biju, V.; Makita, Y.; Nagase, T.; Yamaoka, Y.; Yokoyama, H.; Baba, Y.; Ishikawa, M. *J. Phys. Chem. B* **2005**, *109*, 14350–14355.
78. Feller, W. *An Introduction to Probability Theory and Its Applications*, Vol. 1, 2nd ed.; John Wiley: New York, 1964; pp 74–77 and 327–330 (Chapters 3.4 and 14.7).
79. Xu, H.; Käll, M. *Phys. Rev. Lett.* **2002**, *89*, 246802-1–246802-4.
80. Futamata, M.; Maruyama, Y. *Appl. Phys. B* **2008**, *93*, 117–130.
81. Osborne, M. A.; Balasubramanian, S.; Furey, W. S.; Klenerman, D. *J. Phys. Chem. B* **1998**, *102*, 3160–3167.
82. Schatz, G. C.; van Duyne, R. P. *Surface Sci.* **1980**, *101*, 425–438.
83. García-Vidal, F. J.; Pendry, J. B. *Phys. Rev. Lett.* **1996**, *77*, 1163–1166.
84. Li, K.; Stockman, M. I.; Bergman, D. J. *Phys. Rev. Lett.* **2003**, *91*, 227402-1–227402-4.
85. Sasaki, K.; Tanaka, Y.; Fujiwara, H. *Proceedings of Yamada Conference LXVI* **2012**, OP08-1–OP08-4.
86. Tanaka, Y.; Sanada, A.; Sasaki, K. *Sci. Rep.* **2012**, *2*, 764-1–764-5.
87. Yoshida, K.; Itoh, T.; Biju, V.; Ishikawa, M.; Ozaki, Y. *Phys. Rev. B* **2009**, *79*, 085419-1–085419-6.
88. Chibisov, A. K.; Görner, H. *Chem. Phys. Lett.* **2002**, *357*, 434–439.
89. Slavnova, T. D.; Chibisov, A. K.; Görner, H. *J. Phys. Chem. A* **2002**, *106*, 10985–10990.
90. Kitahama, Y.; Ogawa, A.; Tanaka, Y.; Obeidat, S.; Itoh, T.; Ishikawa, M.; Ozaki, Y. *Chem. Phys. Lett.* **2010**, *493*, 309–313.

Chapter 5

3D SERS Imaging

Sanpon Vantasin and Yukihiro Ozaki*

Department of Chemistry, School of Science and Technology, Kwansei Gakuin University, Sanda, Hyogo 669-1336, Japan
***E-mail: ozaki@kwansei.ac.jp**

Three dimensional surface enhanced Raman spectroscopy imaging is a technique that combines 3D Raman imaging with SERS. In this chapter, the rationale and procedures for 3D SERS are discussed. Several schemes to provide suitable substrates for 3D SERS imaging are presented alongside their applications.

Introduction

Raman imaging is a powerful tool for acquiring molecular information from throughout area of interest. Its true strength is not just in the many spectra obtained from multiple points, but also their correlation. The variations, uniformity, and trends among the spectra can provide information such as the distribution of targeted molecules, homogeneity of the sample, lattice consistency, etc. (*1–3*) This allows Raman imaging to be feasible in the study of nanomaterials (*4–6*), biological samples (*7–10*), polymers (*11–14*), and various other kinds of samples (*15–20*). The vibrational information of the molecules in the area can be correlated with structural information from microscopy techniques to explore the visible and invisible parts in much deeper aspect. For example, typical optical microscope images contain color channel of red, blue, and green. Raman imaging provides hyperspectral images with hundreds of "color" channels in different wavenumbers. Since the images are generated from totally different mechanism to optical images, the two types of images greatly complement each other (*21–23*).

Incorporating the technique of surface-enhanced Raman spectroscopy (SERS), which is well known for its extremely high signal enhancement and near-surface information, into Raman imaging, results in SERS imaging. This technique allows low-concentration molecules to be visualized (*24*, *25*). The localized enhancement of SERS, together with antibody-modified SERS probes can precisely locate certain targets (e.g., cancer cells) and present them on SERS images (*26*, *27*). SERS imaging is also useful for SERS substrates with nonuniform hotspot: by averaging SERS signals form a large area instead of a single point, the signal intensity can be obtained with considerably less selection bias (*28*, *29*).

For decades, three-dimensional optical images have been conventionally generated using confocal microscopy by taking and combining images at various Z depths (*30*, *31*). This is possible due to the original purpose of confocal microscopy, i.e., to acquire light from a narrow designated XY plane. Since Raman imaging instruments often be equipped with a confocal system, the same principle of 3D images construction can be applied to Raman imaging as well. Three-dimensional Raman imaging is useful because real samples are three dimensional, with 3D structures and distributions of chemical substances. The images from 3D Raman imaging are closer to the true nature of the sample than those from 2D Raman imaging. Three-dimensional heterogeneous structures (*11*), distribution of toxic substances in cells (*32*), biochemical composition of organelles in cells (*33*), and strain in 3D crystals (*34*, *35*) have been explored with great details with 3D Raman imaging.

Adding the third dimension to Raman imaging is trivial. Instead of a motorized stage with only two axes of movement, another controller and motor is incorporated to allow three dimensional positioning. Adding the third dimension to SERS imaging is, on the other hand, not trivial. By definition, surface(s) of SERS substrate is required for SERS. Some mechanisms are needed to provide the presence of SERS substrate in three dimensions, while allowing lasers and scattered signal to access the entire imaged volume in order to acquire SERS signal in 3D. Although 3D SERS substrates contain hotspots in three dimensions, most of them are either too opaque or cannot be easily embedded in interesting samples, or both. This will be further discussed in the Part 3 of this chapter. At the time of writing, there are not many publications about 3D SERS imaging. (Although some other manuscripts used the term "3D SERS imaging" or "3D SERS mapping", they actually provided 3D plots of 2D SERS imaging data.). Despite the difficulty of 3D SERS imaging, these studies produced important data that cannot be acquired by either 2D SERS imaging or 3D Raman imaging. They also demonstrated methods that can be used for 3D SERS imaging in other samples. These studies will be discussed in three parts according to the techniques used to provide the SERS hotspots in three dimensions: controlled aggregation of metallic nanoparticles in 3D, 3D tracking of metallic nanoparticles, and hierarchical 3D SERS substrates.

Controlled Aggregation of Metallic Nanoparticles in 3D

McAughtire et al. (*36*) and Chen et al. (*37*) demonstrated controlled aggregation of metallic nanoparticles using eukaryotic cells as supporting structures. Aggregated nanoparticles, the classic SERS hotspot provider, are still the SERS substrate with decent SERS enhancement (*38*, *39*). SERS enhancement factor up to 10^4–10^6 is easily achievable using aggregated nanoparticles (*40*).

While the concentrations of nanoparticles are sufficiently high enough to provide the SERS effect, they are not enough to make the whole cell opaque. This transparency combined with the distribution of nanoparticles aggregating onto the natural 3D structure of cells, allows the SERS signal to be measured throughout the 3D volume. Thus, 3D SERS imaging can be performed.

In the study of McAughtire et al. (*36*), the controlled aggregation was done before the introduction of nanoparticles into the cells. Citrate-capped silver nanoparticles (AgNPs), with an average size of 42 nm, were induced-aggregated with 1,6-hexamethylene diamine; however, the aggregation is not complete because polyvinylpyrrolidone (PVP) was added during the aggregation. PVP is a common stabilizer for AgNPs due to its good affinity with the nanoparticles, water solubility, and polymeric chain providing steric repulsion (*41*). This provides small clusters of AgNPs with a lot of hotspots, while retaining the ability to disperse in solutions. Four lots of aggregated AgNPs were then separately labeled with four small and common Raman probe molecules (4-mercaptopyridine (MPY), 5,5′-dithiobis(2-nitrobenzoic acid) (DTNB), 4-nitrobenzenethiol (NBT), and 2-naphthalenethiol (2-NPT)). In their paper, McAughtire et al. called these labeled AgNPs clusters "nanotags".

The nanotags were then introduced into Chinese hamster ovarian cells. The 3D SERS image acquired with a 633 nm excitation laser (presenting nanotags) was overlaid with the 3D Raman image using a 532 nm laser (presenting cell structures). The result is shown in Figure 1.

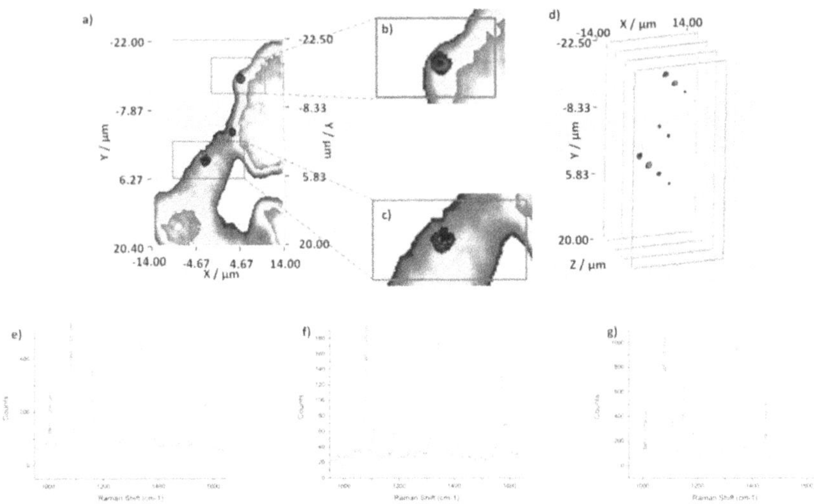

Figure 1. (A) 2D slice overlay at Z = 0.000 µm between 3D SERS imaging of nanotags and 3D Raman imaging of a Chinese hamster ovarian cell. (B),(C) Magnified images of the corresponding areas in (A), showing nanotags. (D) Multiple 2D slices showing nanotag positions in 3D. (E–G) SERS spectra from top, middle, and bottom nanotags clusters, respectively. Reproduced from reference (36). Copyright 2013, PCCP Owner Societies, reprinted under a Creative Commons Attribution 3.0.

As seen in the correspondence between the 3D Raman image of the cell organelle and the 3D SERS image of the nanotags, it is certain that the labeled nanoparticles were incorporated into the cell. 2D SERS images cannot exactly confirm this because the nanotags that appear to be "inside" the cell in a 2D SERS image might actually be on the top or bottom, outside the cell. Electron microscopy can measure the cellular uptake; however, it is a destructive technique. McAughtire et al. suggested the potential applications of 3D SERS imaging with multi-markers in cellular disease detection (*36*).

Chen et al. (*37*) further developed multi-marker 3D SERS imaging to a whole new level. A universal synthesis method for labeled or label-free, membrane- or nucleus-targeting gold nanoparticles (AuNPs) was established in this work. Briefly, 40 nm AuNPs were functionalized by one of the Raman dyes (4-mercaptobenzoic acid, crystal violet (CV), or cresyl violet acetate (CVa)) for labeled probes, or by no dye at all for label-free probes. Then, polyallylamine was added to the surface of the AuNPs to provide –NH2 groups for further modification. Finally, nucleus- or membrane-targeting peptides were conjugated onto the particles by amide coupling to polyallylamine. The functionalized AuNPs were then introduced into HeLa cells via endocytosis. 3D SERS imaging was performed using a 632.8 nm He-Ne laser. The laser spot is around 1 μm wide and the step between each point is also 1 μm. Exposure time for each point is 1 s.

The result is a complete visualization of the cellular structures in 3D. The labeled probes provide the precise location of nucleus and cell membrane with extremely sharp contrast, while the label-free probes provide molecular information, such as proteins and DNA, from the organelles they are attached on. The 3D property is clearly shown in Figure 2, where the cell membrane locations in various Z depths are presented. It is obvious that, unlike 2D SERS imaging, this 3D imaging presents the structural information, chemical information, and activities of the cell in three dimensions.

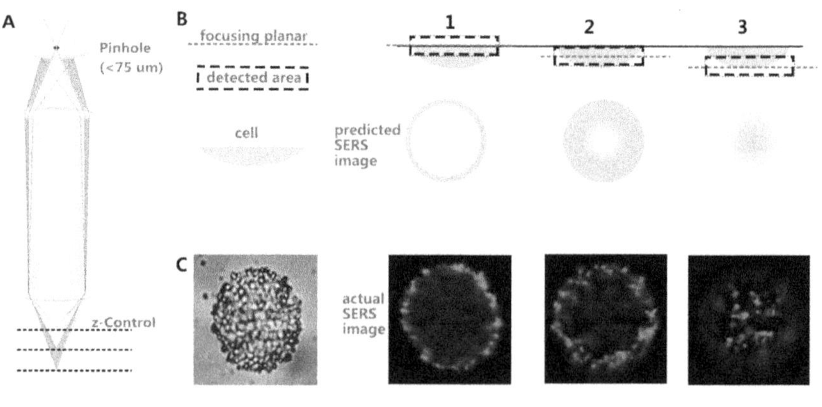

Figure 2. (a) Confocal setup of the 3D SERS measurement in reference (37). (b), (c) Predicted and measured 3D SERS images of labeled AuNPs targeting cell membrane, respectively. The SERS images are XY slices taken at 0, 3, and 6 μm on the Z axis. Reproduced from reference (37). Copyright 2016, Springer Nature, reprinted under Creative Commons Attribution 4.0.

A capability in cellular activity monitoring is demonstrated by using Triton-X to induce apoptosis. Time-dependent 3D SERS imaging was then conducted over 48 hours (Figure 3). The result presents the changes in the proteins and DNA through the stages of apoptosis. Figure 3A shows the decreasing trend of signals from the CV- and CVa-labeled AuNPs over the apoptosis, which are targeting on two membrane proteins. This indicates protein detachment from the membrane surface. In Figure 3B, SERS images using label-free nucleus-targeting AuNPs are displayed. The images are generated from combined intensities at 490 and 1630 cm^{-1}. During apoptosis, the DNA signal gets stronger over time, indicating DNA leakage from the nucleus. After 24 hours, a strong DNA-histone complex signal appears throughout the cytoplasm as a result of nuclear rupture. Further analysis illustrates that apoptosis involves many complicated biochemical processes, including, but not limited to, protein degradation and DNA fragmentation.

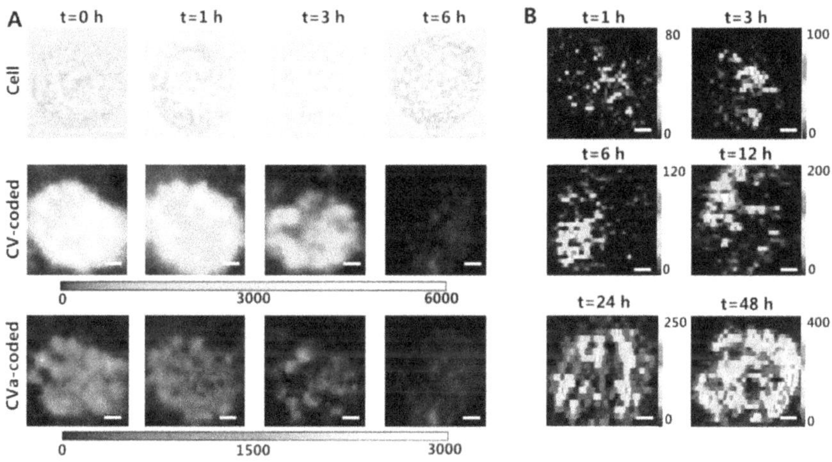

Figure 3. SERS imaging of cell during apoptosis. (A) Optical image and SERS imaging using membrane-targeting labeled AuNPs (B) SERS imaging using label-free nucleus-targeting AuNPs. Reproduced from reference (37). Copyright 2016, Springer Nature, reprinted under Creative Commons Attribution 4.0.

Although this technique of 3D SERS using controlled nanoparticles aggregation has only been used for cells, its applications in other systems are highly possible. For example, three dimensional SERS-active superstructures from self-assembling AgNPs in oil-in-water emulsions have been demonstrated for trace detection of drugs urine (*42*). Such superstructures, with sizes in the microscale range, might be suitable substrates for 3D SERS imaging.

3D Tracking of Metallic Nanoparticles

Although the works of Chen et al. obtained rich chemical information in each part of the cells using SERS spectra from three dimensional volumes, they did not explore another crucial topic: dynamics.

Two papers from the same research group, Huang et al. (*43*) and Bando et al. (*44*), extended the simultaneous nanoparticle tracking and single nanoparticle SERS measurement approach by Ando et al. (*45*) (also from the same group) to three dimensions. This allows real-time study of dynamics of intracellular transportation.

A low concentration of 80 nm AuNPs was introduced into a macrophage/HeLa cell via endocytosis. The reason behind the low concentration is that only one nanoparticle should be followed at a time. A dark field optical image was then taken using a 100 ms high-speed camera, with the cell and nanoparticle illuminated by white light from a halogen lamp. The XY position of the particle is acquired from the captured image through image processing. Using the XY position, a mirror attached to motorized galvanometer adjusted the angle and directed the laser beam through an objective lens onto the particle. The position of the gold particle on the Z axis was tracked by capturing optical images at focused and off-focused plane. The perceived intensities at the particle's XY position between the two images were then compared using a pre-calibrated curve to evaluate the real Z position. A piezo stage was used to adjust the Z position accordingly (Figure 4). The Raman signal was continuously collected via the same objective lens and detected by a spectrometer, with 50–100 ms exposure time per frame. The accuracy of tracking is 13 nm in the X and Y axes and 33 nm for the Z axis. The spatial resolution of the Raman signal is ~93 nm for the X and Y axes and ~113 nm for the Z axis.

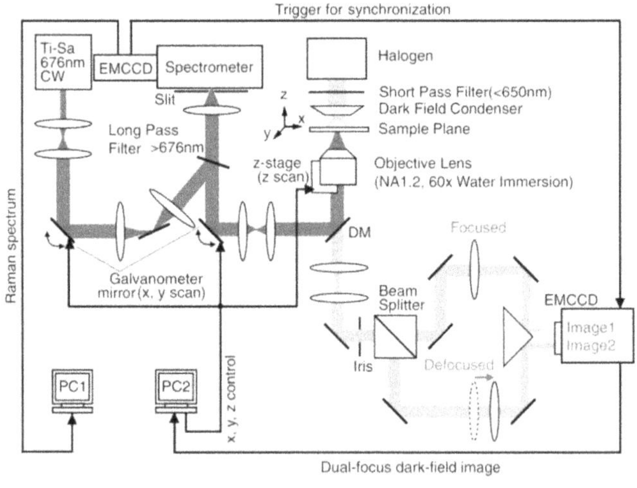

Figure 4. Optics setup for 3D tracking of gold nanoparticle with SERS measurements. Reproduced from reference (43). Copyright 2014, Elsevier.

As both the position and chemical information were tracked throughout the 50 seconds of measurement, the behavior of endocytosed AuNPs can be studied in great details. In the work of Huang et al., the movement of the particle consisted of periods of random motion and relatively straightforward motion. The SERS spectra during periods of random motion presented relatively strong intensities at 1456 cm^{-1} ($C–H_2$ and $C–H_3$) and 1582 cm^{-1} (phenylalanine or tyrosine).

Meanwhile, the SERS spectra from the periods of straightforward motion presented much stronger intensities at 1106 cm^{-1} (C–N stretching), representing proteins. Huang et al. suggested that the protein signal should indicate that in straightforward motions, transportation proteins (e.g., kinesin or dynein) are involved, and thus the SERS signal from the proteins appears (Figure 5).

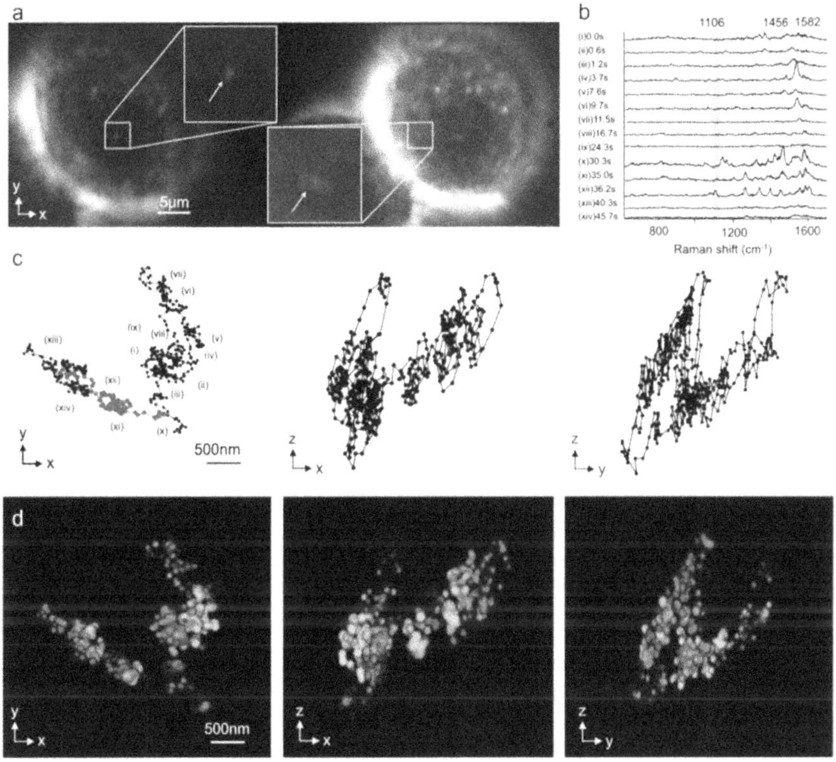

Figure 5. (A) Dark field optical image of an AuNP in a macrophage cell. (B) SERS spectra from the nanoparticle at different times throughout the experiment. (C) Recorded positions of the AuNP viewed from Z, Y, and X directions, respectively. (D) 3D maps generated from the intensities of 1106 cm^{-1}, 1456 cm^{-1}, and 1582 cm^{-1} bands, viewed from the corresponding directions in (C). Reproduced from reference (43). Copyright 2014, Elsevier.

Bando et al. presented a similar experiment on HeLa cells, but with some chemometrics techniques, such as principal component analysis (PCA) and 2D correlation, to analyze the measured SERS spectra. Using PCA, principal component number 3 (PC3), which represents ribose, phosphate, nucleobases, and amide III bands, can be associated with straightforward motion. PC11, which represents ribose phosphate, phenylalanine, amide III, and nucleobases, is associated with random motion. 2D correlation shows that in the first random motion period, the nucleobases or amide III band at 1294 cm^{-1} is strongly correlated with the nucleotides band at 1384 cm^{-1}. The straightforward motion period shows a correlation between the phosphate band at 1090 cm^{-1} and the lipid/protein band at 1448 cm^{-1}. The relationship between particle movement speeds and SERS spectra was also investigated. Most noticeably, the 1084 cm^{-1} band (histidine/phosphate) usually appeared before the particle changed speed. Bando et al. suggested that this might indicate ATP, which might be used as a signal to drive the transportation. With these results, more information about intracellular transportation has been revealed.

Tracking the position and continuously measuring SERS signal means that each SERS spectrum has a corresponding XYZ-position and time. Therefore, in some sense, this process might be considered as a four-dimensional SERS imaging. Another viewpoint is that at each point in time, there is only one spectrum measured from one point in XYZ space. The "mapping" from this process can be also considered a one-dimensional imaging spanning four-dimensional space. Semantics aside, this technique is undoubtedly useful for the study of complicated processes within cells.

Both the controlled aggregation and 3D tracking schemes have been mainly focused in cell applications, perhaps because the benefits of understanding cells are immense. However, this should not be a limitation of how these techniques are applied. There should be some other application as well, such as the internal study of cell-like frameworks of solids in liquids, or emulsion systems.

However, some drawbacks of the two schemes for 3D SERS imaging are their stability and controllability. For the controlled aggregation method, the degree of control is not perfect as the aggregation cannot be made fully regular (i.e., some randomness exists in the aggregation). The signal intensity, which is dependent on aggregation shape and size, is therefore not so qualitative. For the single nanoparticle tracking method, the movement of particle cannot be controlled. Therefore, some interesting spots cannot be probed because the particle does not move to those points. A more stable scheme is needed for 3D SERS imaging in material science. The solution is to use a free-standing hierarchical 3D SERS substrate for 3D SERS imaging.

Hierarchical 3D SERS Substrate for 3D SERS Imaging

Two papers from Kodiyath et al. (*46*, *47*) demonstrated an experiment that can be considered to be the first 3D SERS imaging of a 3D SERS substrate. XY mappings of the SERS signal from AgNPs-decorated porous alumina nanocylinder arrays were collected at various Z depths. However, the phrases "3D SERS imaging" and "3D SERS mapping" do not appear at all in the two papers. Perhaps, even though 3D SERS imaging is quite useful in the characterization of this substrate, it would not provide more information than depth profile SERS in the Z axis. In any case, 3D SERS imaging has never been the main focus of the two studies, as the strengths of the studied substrate are the strong SERS signal, uniformity of enhancement, and large surface area. This is sufficient to find broad applications in trace molecular detection.

Zhang et al. (*48*) made a breakthrough in 3D SERS imaging by making silver-coated 3D micropyramids decorated with silver nanocubes. This was the first time that SERS signals could be acquired from entire stable three-dimensional structures with high signal uniformity. The pyramid shape allows the signal from the bottom part to be observed with little blocking from the top part.

The substrate was synthesized by using 3D laser lithography on a photoresist. The fabrication was controlled by an XYZ piezo stage to control the polymerization spot precisely. Silver film of 50 nm thickness was then coated over the lithographed pyramid structure. 4-methylbenezenthiol (4-MBT) was deposited on the silver film, acting as SERS probe molecules. Since the film is smooth, it does not provide strong SERS enhancement. Therefore, silver nanocubes were deposited on the silver film through Langmuir–Blodgett assembly method. In this preparation, 4-MBT is "sandwiched" between the silver film and silver nanocubes; thus, its SERS signal is strongly enhanced. 3D SERS imaging was done with a 532 nm excitation laser, 2.2 μW of laser power, 10 seconds of exposure time per point, and 0.5 μm step sizes.

The SERS signal from the micropyramid is strong and very predictable. The stability is also high: as the substrate is a free standing solid, a second or third scan on the same structure would yield very similar results. Various shapes and sizes of micropyramids were studied: The size of pyramids does not significantly affect the enhancement factor as the enhancement is from the small nanoscale structure, not the overall shape of the large structure. Other shapes, such as a steep pyramid, a right prism, a truncated cone, and a square block were are also studied. They all showed reasonable 3D SERS images corresponding to the 3D structures and strong enhancement, except for at the points near the sharp slope or perpendicular step (which would result in SERS signal blocking similar to that in the work of Kodiyath et al.).

Decorated micropyramids of various shapes can be used for 3D encodings of digital data, as shown in Figure 6. An advantage of 3D encodings is apparent, as drastically more data can be encoded compared to 2D encodings with the same area. The drawing in Figure 6 represents two structures that would be decoded into the same data in 2D encodings: 3 tracks between 4 pitches. In 3D, with SERS imaging, the two structures can be decoded into totally different vale (Figure 6B vs 6E, and 6C vs 6F).

Although Zhang et al. only demonstrated 3D SERS imaging of sandwiched 4-MBT molecules. This 3D substrate should be able to provide 3D SERS imaging of other systems as well, such as in the solution with a concentration gradient of target molecules in the Z-axis. However, the SERS enhancement might be low because the "sandwich" effect would not be presented as in the case of 4-MBT.

To encourage broader application of 3D SERS imaging, we (Vantasin et al.) developed highly symmetric nanoporous silver microparticles for 3D SERS imaging (*49*). Unlike the substrate of Zhang et al., our substrate is purely chemically synthesized and dose not bond to solid substrates. The microparticles can be embedded in any kind of solution/polymer that does not react with silver. The microparticles have intrinsic nanopores and SERS activity. Therefore, they do not need any nanoparticle decoration. This is useful because it reduces the risk of nanoparticle detachment when embedded into solution/polymer systems. Since the particles are not bonded to solid substrate, they can be easily dispersed throughout the sample to allow 3D SERS imaging anywhere in the sample, and not just the bottom part.

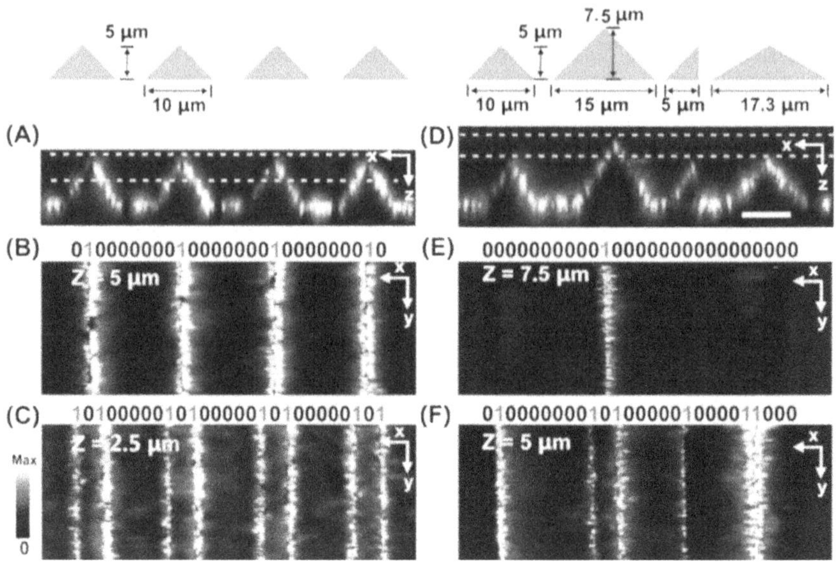

Figure 6. Schematic drawing represents micropyramidal structures and their 3D SERS imaging result. (A),(D) XZ slices of the two structure presenting side-view 3D SERS images. (B),(C) XY slices at the labeled Z depths in 3D SERS imaging of the first structure. (E),(F) Similar XY slices from the second structure. Reproduced from reference (48). Copyright 2014, John Wiley & Sons, Inc.

Hexapod AgCl microparticles were synthesized by controlled precipitation reaction of $[Ag(NH_3)_2]^+$ by NaCl. The resulting particle has a highly symmetric shape (octahedral symmetry) due to the kinetics and thermodynamics of crystal growth. After being washed, the AgCl particles were converted into Ag particles with an in-place galvanic reaction using a Zn plate as the reducer and 0.1 M NaCl as the electrolyte. SEM images show uniform nanopores with a size of 60 nm.

Using *p*-aminothiophenol (PATP) as the probe molecule, 3D SERS patterns of the silver nanoporous microparticles are evaluated (Figure 7). The concentration of PATP on the silver surface should be uniform, because PATP is adsorbed on silver as a monolayer and unadsorbed PATP is washed out by ethanol. Therefore, the peak area mapping shown in Figure 7 also represents the enhancement pattern in 3D. It is clear that the enhancement pattern is very symmetric and predictable, as it closely resembles the hexapod shape of the particles. This point is crucial when using these nanoporous microparticles to study real-life samples. Any irregularities in the peak area mapping must come from the inhomogeneity of the sample, as the enhancement pattern is very predictable.

An interesting point is that the concentration of the particles does not affect the 3D SERS image at all as the 3D SERS imaging is performed on a single microparticle. Therefore, arbitrarily low concentrations of microparticles can be used when the bulk physical properties of the sample need to be kept as is.

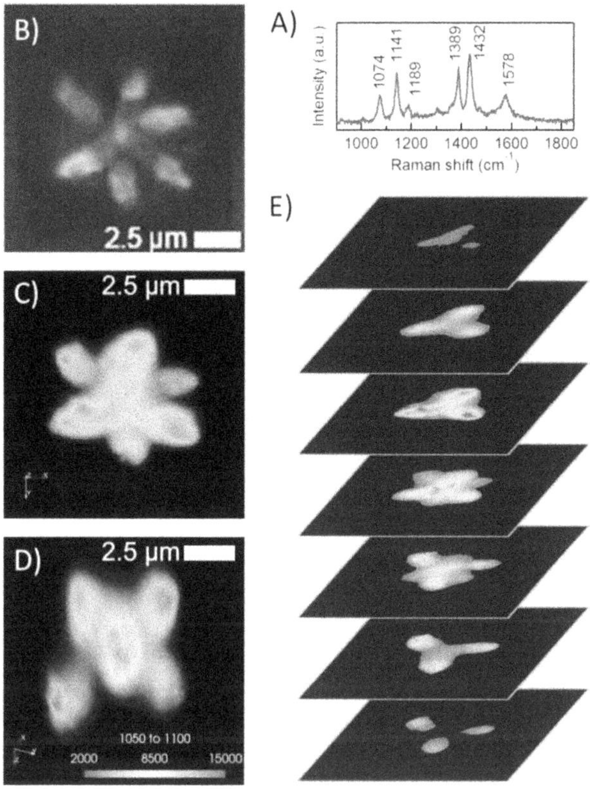

Figure 7. (A) SERS spectra of PATP (B) Optical microscopic image of a hexapod nanoporous silver microparticle. (C–E) Top view, side view, and XY slices of the 3D SERS image acquired from the particle, generated using peak area of the C–S stretching peak at 1074 cm^{-1}. Reproduced from reference (49). Copyright 2016, John Wiley & Sons, Inc.

The utility of this 3D SERS substrate is demonstrated by embedding the particles in a polystyrene/polyvinylpyrrolidone (PS/PVP) 2-layer polymeric system. Such a system can be characterized with 3D Raman imaging; however, the refraction at the polymer/polymer and polymer/air interfaces greatly smears out the laser spot. The laser spot can be as large as 22 µm in the Z-axis when probing deeper into the polymer (*50, 51*). This drastically worsens the spatial resolution in the Z-axis, even with a confocal system (*50, 51*). Conventional 3D Raman imaging cannot distinguish the PS and PVP layers well in the experiment. On the other hands, 3D SERS imaging using symmetric nanoporous silver microparticle determines the transition from PS-like spectra to PVP-like spectra within just 1.2 µm. This is due to the confinement of the probing volume: the probing volume in SERS is only near the silver surface, and thus, only a small volume at the designated Z depth is probed even though the laser spot is smeared out by refraction. In the paper, we described this effect as an improvement in spatial resolution in the Z-axis. However, the more accurate explanation for the higher distinguishing power between different Z depths is that: the 3D symmetric microparticles allows the measurement of SERS signals in a small volume at the desired Z depth, by choosing the correct XY position corresponding to the Z value. Evaluating the XY position from the Z position, and vice versa, is a trivial task with the octahedrally symmetric shape.

Conclusion

In conclusion, we discussed the three schemes for SERS substrate preparation, 3D SERS imaging on the substrates, and their current and potential applications. Although 3D SERS imaging is relatively new, this technique has strong potential for use in the in-depth characterization and visualization of many real-life systems.

References

1. Stewart, S.; Priore, R. J.; Nelson, M. P.; Treado, P. J. *Annu. Rev. Anal. Chem.* **2012**, *5*, 337–360.
2. Opilik, L.; Schmid, T.; Zenobi, R. *Annu. Rev. Anal. Chem.* **2013**, *6*, 379–398.
3. Salzer, R.; Siesler, H. W. *Infrared and Raman Spectroscopic Imaging*; Wiley-VCH: Weinheim, 2009.
4. Ajito, K.; Sukamto, J. P. H.; Nagahara, L. A.; Hashimoto, K.; Fujishima, A. *J. Vac. Sci. Technol. A* **1995**, *13*, 1234–1238.
5. Ni, Z. H.; Yu, T.; Lu, Y. H.; Wang, Y. Y.; Feng, Y. P.; Shen, Z. X. *ACS Nano* **2008**, *2*, 2301–2305.
6. Larsen, R. A.; Kubo, Y.; Akao, K.; Ohkubo, Y.; Yumoto, M.; Nagoshi, T. *AIP Conf. Proc.* **2010**, *1267*, 766–767.
7. Liu, Z.; Li, X.; Tabakman, S. M.; Jiang, K.; Fan, S.; Dai, H. *J. Am. Chem. Soc.* **2008**, *130*, 13540–13541.
8. Okada, M.; Smith, N. I.; Palonpon, A. F.; Endo, H.; Kawata, S.; Sodeoka, M.; Fujita, K. *Proc. Natl. Acad. Sci.* **2012**, *109*, 28–32.
9. Keren, S.; Zavaleta, C.; Cheng, Z.; Zerda, A. de la; Gheysens, O.; Gambhir, S. S. *Proc. Natl. Acad. Sci.* **2008**, *105*, 5844–5849.
10. van Apeldoorn, A. A.; van Manen, H.-J.; Bezemer, J. M.; de Bruijn, J. D.; van Blitterswijk, C. A.; Otto, C. *J. Am. Chem. Soc.* **2004**, *126*, 13226–13227.
11. Smitthipong, W.; Gadiou, R.; Vidal, L.; Wagner, P.; Nardin, M. *Vib. Spectrosc.* **2008**, *46*, 8–13.
12. Markwort, L.; Kip, B. *J. Appl. Polym. Sci.* **1996**, *61*, 231–254.
13. Schmidt, U.; Liu, W.; Yang, J.; Dieing, T.; Weishaupt, K. *Microsc. Microanal.* **2015**, *21*, 2189–2190.
14. Schaeberle, M. D.; Karakatsanis, C. G.; Lau, C. J.; Treado, P. J. *Anal. Chem.* **1995**, *67*, 4316–4321.
15. Fateixa, S.; Wilhelm, M.; Nogueira, H. I. S.; Trindade, T. *J. Raman Spectrosc.* **2016**, *47*, 1239–1246.
16. *Confocal Raman Microscopy*; Dieing, T., Hollricher, O., Toporski, J., Eds.; Springer Series in Optical Sciences; Springer: Heidelberg, Germany; New York, 2010.

17. Clarke, F. C.; Jamieson, M. J.; Clark, D. A.; Hammond, S. V.; Jee, R. D.; Moffat, A. C. *Anal. Chem.* **2001**, *73*, 2213–2220.
18. Gordon, K. C.; McGoverin, C. M. *Int. J. Pharm.* **2011**, *417*, 151–162.
19. Ropret, P.; Miliani, C.; Centeno, S. A.; Tavzes, Č.; Rosi, F. *J. Raman Spectrosc.* **2010**, *41*, 1462–1467.
20. Neff, D.; Bellot-Gurlet, L.; Dillmann, P.; Reguer, S.; Legrand, L. *J. Raman Spectrosc.* **2006**, *37*, 1228–1237.
21. *Raman Imaging: Techniques and Applications*; Zoubir, A., Ed.; Springer Series in Optical Sciences; Springer: Heidelberg, Germany; New York, 2012.
22. Christensen, K. A., *Hyperspectral and Filter-Based Raman Microscopy and Imaging: Design and Evaluation of Practical Imaging Systems*, Ph.D. thesis, University of Michigan, Ann Arbor, MI, 1997.
23. Sasic, S.; Ozaki, Y. *Raman, Infrared, and Near-Infrared Chemical Imaging*; John Wiley & Sons: Hoboken, NJ, 2011.
24. Ishikawa, M.; Maruyama, Y.; Ye, J. Y.; Futamata, M. *J. Lumin.* **2002**, *98*, 81–89.
25. Fan, Z.; Senapati, D.; Khan, S. A.; Singh, A. K.; Hamme, A.; Yust, B.; Sardar, D.; Ray, P. C. *Chem. – Eur. J.* **2013**, *19*, 2839–2847.
26. Lee, S.; Chon, H.; Yoon, S.-Y.; Lee, E. K.; Chang, S.-I.; Lim, D. W.; Choo, J. *Nanoscale* **2012**, *4*, 124–129.
27. Yigit, M. V.; Medarova, Z. *Am. J. Nucl. Med. Mol. Imaging* **2012**, *2*, 232–241.
28. Zhou, M.; Liu, X.; Yu, B.; Cai, J.; Liao, C.; Ni, Z.; Zhang, Z.; Ren, Z.; Bai, J.; Fan, H. *Appl. Surf. Sci.* **2015**, *333*, 78–85.
29. Hwang, J.; Lee, S.; Choo, J. *Nanoscale* **2016**, *8*, 11418–11425.
30. Sheppard, C. J. R.; Cogswell, C. J. *J. Microsc.* **1990**, *159*, 179–194.
31. Schmidt, U.; Yang, J.; Liu, W.; Jauss, A.; Dieing, T. *Microsc. Microanal.* **2011**, *17*, 1196–1197.
32. Seydou, Y.; Chen, H.-H.; Harte, E.; Ventura, G. D.; Petibois, C. *Anal. Bioanal. Chem.* **2013**, *405*, 8701–8707.
33. Majzner, K.; Kaczor, A.; Kachamakova-Trojanowska, N.; Fedorowicz, A.; Chlopicki, S.; Baranska, M. *Analyst* **2013**, *138*, 603–610.
34. del Corro, E.; Izquierdo, J. G.; González, J.; Taravillo, M.; Baonza, V. G. *J. Raman Spectrosc.* **2013**, *44*, 758–762.
35. Crisci, A.; Baillet, F.; Mermoux, M.; Bogdan, G.; Nesládek, M.; Haenen, K. *Phys. Status Solidi A* **2011**, *208*, 2038–2044.
36. McAughtrie, S.; Lau, K.; Faulds, K.; Graham, D. *Chem. Sci.* **2013**, *4*, 3566.
37. Chen, Y.; Bai, X.; Su, L.; Du, Z.; Shen, A.; Materny, A.; Hu, J. *Sci. Rep.* **2016**, *6*.
38. Ko, H.; Singamaneni, S.; Tsukruk, V. V. *Small* **2008**, *4*, 1576–1599.
39. Sharma, B.; Cardinal, M. F.; Kleinman, S. L.; Greeneltch, N. G.; Frontiera, R. R.; Blaber, M. G.; Schatz, G. C.; Duyne, R. P. V. *MRS Bull.* **2013**, *38*, 615–624.
40. Cañamares, M. V.; Garcia-Ramos, J. V.; Sanchez-Cortes, S.; Castillejo, M.; Oujja, M. *J. Colloid Interface Sci.* **2008**, *326*, 103–109.
41. El Badawy, A. M.; Scheckel, K. G.; Suidan, M.; Tolaymat, T. *Sci. Total Environ.* **2012**, *429*, 325–331.

42. Han, Z.; Liu, H.; Wang, B.; Weng, S.; Yang, L.; Liu, J. *Anal. Chem.* **2015**, *87*, 4821–4828.
43. Huang, K.-C.; Bando, K.; Ando, J.; Smith, N. I.; Fujita, K.; Kawata, S. *Methods* **2014**, *68*, 348–353.
44. Bando, K.; Smith, N. I.; Ando, J.; Fujita, K.; Kawata, S. *J. Opt.* **2015**, *17*, 114023.
45. Ando, J.; Fujita, K.; Smith, N. I.; Kawata, S. *Nano Lett.* **2011**, *11*, 5344–5348.
46. Kodiyath, R.; Wang, J.; Combs, Z. A.; Chang, S.; Gupta, M. K.; Anderson, K. D.; Brown, R. J. C.; Tsukruk, V. V. *Small* **2011**, *7*, 3452–3457.
47. Kodiyath, R.; Papadopoulos, T. A.; Wang, J.; Combs, Z. A.; Li, H.; Brown, R. J. C.; Brédas, J.-L.; Tsukruk, V. V. *J. Phys. Chem. C* **2012**, *116*, 13917–13927.
48. Zhang, Q.; Lee, Y. H.; Phang, I. Y.; Lee, C. K.; Ling, X. Y. *Small* **2014**, *10*, 2703–2711.
49. Vantasin, S.; Ji, W.; Tanaka, Y.; Kitahama, Y.; Wang, M.; Wongravee, K.; Gatemala, H.; Ekgasit, S.; Ozaki, Y. *Angew. Chem. Int. Ed.* **2016**, *55*, 8391–8395.
50. Everall, N. J. *Appl. Spectrosc.* **2000**, *54*, 1515–1520.
51. Everall, N. J. *Appl. Spectrosc.* **2000**, *54*, 773–782.

Chapter 6

Exploring the Effect of Intermolecular Hydrogen Bonding and the Application in Label-Free Enantioselective Discrimination by SERS

Yue Wang,[1] Bing Zhao,[*,1] and Yukihiro Ozaki[2]

[1]State Key Laboratory of Supramolecular Structure and Materials, Jilin University, Changchun 130012, P. R. China
[2]Department of Chemistry, School of Science and Technology, Kwansei Gakuin University, Sanda, Hyogo 669-1337, Japan
[*]E-mail: zhaob@mail.jlu.edu.cn. Fax: +86 431 8519 3421. Tel: +86 431 8516 8473.

Surface-enhance Raman scattering (SERS) spectroscopy has been verified to be a powerful technique among current methods for the investigations of hydrogen bonding. SERS offers rich chemical and structural information about molecules, and has single-molecule sensitivity, which makes it a remarkable analysis technique in very wide areas of science. In this chapter, a SERS method for exploring the influences of intermolecular hydrogen bonding was demonstrated. It was found that significant changes occurred both in the vibrational frequencies and intensities of molecules in a hydrogen bonding assembled system due to the modifications in their electronic structure under the influence of hydrogen bonding. In virtue of the selective enhancement in the SERS spectra, a conclusion can be drawn that intermolecular hydrogen bonding promotes the charge transfer transition between a substrate and an adsorbate in an assembled system. Furthermore, a label-free approach for enantioselective discrimination of chiral alcohols by SERS was thus proposed on the basis of the effect of hydrogen bonding, in which the relative intensities of the achiral selector molecule was associated with the chiral environment.

Introduction

Hydrogen bonding is a fundamentally important force in nature, which is mediated through the interaction between a proton donor and an electronegative proton accepter (*1*, *2*). It has been widely observed in various fields of science, such as biological, chemical, and material sciences. There are two kinds of hydrogen bonding: intramolecular and intermolecular ones. The former hydrogen bonding is used to serve to explain various molecular properties, and can be responsible for the geometry of a certain conformation or molecule (*3*, *4*). On the other hand, the latter hydrogen bonding is crucial for molecular aggregation, molecular assemblies, and properties of hydrogen-bond-based supramolecular materials (*5*, *6*). It plays a key role not only in the function of biomolecules in living organisms, including biological electron transfer, enzyme activity, biometric recognition and many other life activities, but also in many molecular and supramolecular systems as diverse as hydrogen-bond crystal engineering, polymers, self-assembled supramolecular architectures, molecular recognition of organic molecules (*5–8*). Although hydrogen bonding has already been investigated for a century, it still remains to be actively studied because of its ubiquity as well as its importance in nature.

In general, hydrogen bonds are weaker than covalent bonds, but stronger than other intermolecular interactions, such as van der Waals force and dispersion force. The formation of a hydrogen bond can result in changes in the electronic structure of a molecule, but the rearrangements arise from hydrogen bonding are much more subtle than the massive shifts of electron density caused by covalent bonds. Therefore, revealing subtle changes of the hydrogen bonding at the molecular level and its influence on the electronic structures of molecules is challenging. Up to now many experimental techniques have been developed for hydrogen bond research. Spectroscopic methods (*9–12*), including infrared (*9*), near-infrared (*10*), Raman (*11*) and terahertz spectroscopy (*12*), are major tools of studying the presence, strength and structure of hydrogen bonding, due to their superiority in nondestructive analyses and in situ test, and they have engendered great research enthusiasm. It can provide information about local molecular geometries and their interaction with the environment. In addition, X-ray diffraction (*13*), neutron diffraction (*14*), electron diffraction (*15*), nuclear magnetic resonance measurement (NMR) (*16*), and theoretical calculation (*17*) have been used for investigating the structure of hydrogen bonds in a molecular system as well.

Among these conventional methods, the rationale and inherent properties of Raman spectroscopy allow it to become a powerful technique for the investigations of intermolecular hydrogen bonding (*11*, *18*, *19*). It has unique advantages over electronic spectroscopy and other techniques, considering that it can provide abundant vibrational modes information correlated to specific vibrational motions of molecules under various environmental conditions (temperature and pressure etc.) (*20*, *21*). In particular, it can offer refined fingerprint region with a wealth of structural information included. By identification of these vibration modes, conclusions can be drawn on specific structures or functional groups in the molecules. Nevertheless, the main drawback of Raman spectroscopy is its weak signal, which results in the lack of enough sensitivity in exploring the influence of hydrogen bonding on the vibration modes of molecule. It would, therefore, be of particular importance to develop a technique that allows one to monitor intermolecular hydrogen bonding in a more efficient and sensitive manner to get a deeper understanding of the effects of hydrogen bonds in the system. In comparison with normal Raman spectroscopy, Surface-enhanced Raman spectroscopy (SERS) enjoys both the advantage of the innate property of Raman spectroscopy and the sensitivity for single-molecular-level detection (*22–24*).

Vibrational Spectroscopy Methods for Studying Hydrogen Bonding

There have been different methods to be used for exploring the properties and structures of the hydrogen bonding in theoretically and experimentally so far, most of which primarily depend on the small shift of electron density from the proton acceptor to the donor with the formation of a hydrogen bond. Take the IR spectroscopy (*9*, *25*) and NMR (*16*, *26*, *27*) methods for instance, the hydrogen bonding can be verified mainly by the changes in the vibrational frequency of the molecules in terms of the former technique, and the alternations in chemical shift for the latter one. In fact, there are no given characteristics for the demonstration of a hydrogen bonding, the more information from the experiment we get is more helpful in the investigation and understanding of hydrogen bonding in system. Of all the methods of the studies of hydrogen bonding, vibrational spectroscopy shows unique advantages, and many researchers have carried out successful studies of hydrogen bonding using these methods.

For vibrational spectroscopy, the information on the structure and the property of hydrogen bonds is derived from relative vibrational bands in spectra. And the formation of hydrogen bonds causes a change in the frequency of a stretching mode of *X*-H group in a hydrogen bond donor, and this change is related to the strength of the hydrogen bond. It may occur as a consequence of an elongation of the *X*-H bond and a large red-shift in vibrational frequency with the new band significantly increased and broadened, which often happens in IR spectra (*28*, *29*). On the other hand, sometimes the formation of hydrogen bonding results in the shortening of *X*-H bond, and accompanied with a blue-shift in vibrational frequency as well as reduced intensity.

IR spectroscopy has its own shortcomings in hydrogen bonding research. For instance, some bonds (for example, υ (OH)) related to hydrogen bonding are very broad in the case of the hydrogen bonding in alcohol system or aqueous system (*25*), which can be explained by a proton fluctuation mechanism and strong interactions with the environment. Thus the spectral resolution of an IR spectrum is limited, and the effective information about *X-H* bond are difficult to be extracted. Besides, we mainly investigate the bands in high frequency and fingerprint regions (above 400 cm^{-1}) in IR spectra. It is difficult to get the effect of hydrogen bonding on the vibration in the low wave-number range. However, Raman spectroscopy display a unique advantage to study hydrogen bonding. The Raman effect can be explicated with the consideration of molecular polarizability

(*30*, *31*). In other words, Raman spectroscopy has a nice response to the change in molecular polarizability. Polarizability describes the behavior of the electron cloud of the molecule, which can be directly influenced by hydrogen bonding. Moreover, Raman spectroscopy can afford refined fingerprint of a molecule with abundant structural information, even in lower wave number range (10-200 cm^{-1}) (*32*). Therefore, it has become a powerful method for the research of hydrogen bonding.

SERS Spectroscopy

SERS has proven to be a powerful technique in various fields of science, including surface and interface science, material science, environmental science, bioscience, and medical diagnosis (*33*, *34*). It is primarily due to the tremendous enhancement to the Raman scattering signal while simultaneously retaining the characteristics of affording abundant vibrational information by Raman spectroscopy. That is, SERS provides the similar vibration information of molecules that normal Raman spectroscopy dose, as well as the single-molecular level sensitivity to the subtle change in molecular polarizability (*22–24*). Thus, it shows great potential in the study of the effect of weak intermolecular hydrogen bonding on the electron structure of molecule in the system.

Enhancement Mechanism of SERS Spectroscopy

The origin of this surface enhanced effect has been debated over the years, but there is now a broad consensus that two major mechanisms contribute to the SERS signal: electromagnetic enhancement and chemical enhancement (*35–40*). To detailed understand the extremely high sensitivity of SERS, large number of researches were published to reveal this amazing enhancement phenomenon. For an enhanced Raman band in SERS spectra, it cannot be interpreted from the perspective of either of the contributions alone. Generally, the two mechanisms work together in concert to produce the overall SERS effect, which amplify the Raman signals of adsorbed resonant molecules as large as 10^{14} orders of magnitude under suitable conditions. One is the long-range electromagnetic enhancement (EM) mechanism, arising from the localized surface plasmons resonance caused by the collective oscillations of the conduction electrons on the surface of metal (*35*, *36*), which is responsible for the most important contribution of the Raman signal. It is closely associated with the nature of nanosized noble metal, the dielectric constant of the surrounding medium, and the distance between the adsorbates and the surface of the metal (*41*). The other multiplicative contributions to the SERS enhancements relies on the modification in the electronic structure of molecules chemically adsorbed onto the roughed metal surface to form molecule–metal complexes, which are generally grouped as chemical enhancement (CM) (*37–40*). CM, as an additional contribution to SERS enhancement, is much smaller than the EM effect, and it can only increase the SERS intensity by the order of $10–10^3$. In spite of this, CM significantly influences the pattern of the SERS spectra either on the frequency shift or relative intensity of the spectral bands. As referred in the literature (*42*), it is commonly regarded there are three kinds of chemical enhancements:

(1) Enhancement based on ground state chemical interactions between the molecule and nanoparticle rather than any excited state in the molecule-metal system.
(2) Resonance Raman enhancement due to the excitation wavelength being resonance with the transition between the ground state and the excited electronic state of the molecule.
(3) Charge transfer (CT) resonance enhancement associated with the resonance excitation of a charge-transfer state between the molecule and the metal nanoparticle.

Among these contributions, CT resonance, which is considered to be a resonant Raman-like process, is the main effect of CM mechanism in a SERS spectrum (*43*, *44*). According to the Albrecht's resonance Raman theory, which originally contains the coupling between the electronic states through nuclear motions, three terms are involved to contribute to the significant enhancement of Raman signal, which are the A term (Franck-Condon term), B term and C term (Herzberg–Teller terms), respectively. For the Franck–Condon term, it yields only totally symmetric vibrations; however, the Herzberg-Teller terms yield both totally and non-totally symmetric vibrations (*44*, *45*). Generally, the spectral changes in SERS can be explained using the Herzberg-Teller terms. Afterwards, Lombardi *et al.* proposed a unified expression for SERS spectroscopy based on the theory of Albrecht, in which CT resonance can be considered involving the transition of an electron from the Fermi level of the metal to an unoccupied molecular orbital of the adsorbate or vice versa (*37*, *46*). The occurrence of CT resonance depends on the fact whether the photon energy of the excitation matches the energy separation between the Fermi level of the metal and the molecular orbital of the adsorbate. Consequently, it greatly influence the relative intensities of different vibrational modes and the vibrational frequency of an adsorbate in a SERS spectrum. In turn, the variation in the SERS pattern may be empirically considered as a manifestation for a change in the CT process in system caused by the change of the local chemical environment.

Exploring Effect of Intermolecular Hydrogen Bonding on Molecular Electronic Structures

As mentioned above, intermolecular hydrogen bonds influence the electronic structure of molecules, and SERS can reflect the changes in the polarization of the molecules efficiently. We introduced SERS spectroscopy to the study of intermolecular hydrogen bonding in the assembled systems. It is found that changes appeared in the SERS spectra of the molecules both in frequency shifts and the relative intensities. Furthermore, there is an obvious positive correlation between these changes and the concentration of the hydrogen bonding ligand in system. In this section, a hydrogen bonding assembled system was introduced to be investigated by using SERS spectroscopy. It is verified that SERS possesses the feasibility to explore intermolecular hydrogen bonding.

SERS Spectra of Hydrogen Bonding Assembled System

The intermolecular hydrogen bonding assembled system was prepared by a self-assembled method, in which a layer of Ag NPs produced following the classical Lee's method (*47*) were assembled on the glass substrates surface by electrostatic interaction. Various SERS-active molecules were introduced in the system via chemical saturation adsorption with the formation of Au-S bonds. Then, the corresponding hydrogen bonding ligand molecules were chosen to construct the hydrogen bonding system, as illustrated in Figure 1. This hydrogen bonding assembled system efficiently avoid the interference of other uncertainties to the SERS spectra, such as the orientation of SERS-active molecules, the aggregations of Ag NPs, and the diversity of components in the fluid system.

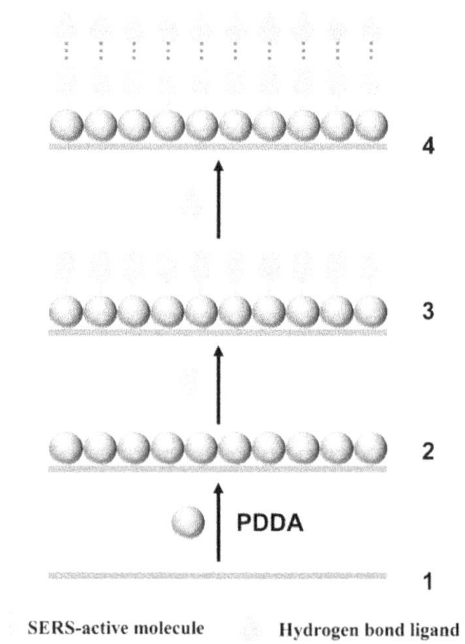

Figure 1. Illustration of the fabrication of intermolecular hydrogen bonding assembled system.

Herein, *p*-mercaptobenzoic acid (MBA), the SERS-active molecule, and aniline, the corresponding hydrogen bonding ligand, were selected to fabricate the intermolecular hydrogen bonding assembled system (see Reference 48 for details). It is found that the SERS pattern of MBA molecules changed significantly with the formation of intermolecular hydrogen bonding in system compared the MBA molecules in blank with the excitation at 633 nm, as shown in Figure 2. The relative intensities of some characteristic bands changed with the introduction of aniline in the system (*48–50*), especially the bands at 417 cm^{-1}, assigned to the

ν(C-S) mode, at 998 and 1022 cm^{-1}, ascribed to the in-plane ring breathing modes, and at 1572 and 1584 cm^{-1}, attributed to the non-totally symmetric ν(C=C) and totally symmetric ν(C=C) mode, respectively. The changes in the spectrum can be excluded the simple superposition of the SERS spectra of MBA and aniline, since the SERS response of aniline is much weaker than that of MBA under same conditions. Considering the only additive in the system is the aniline molecules, the changes in the SERS spectra of MBA is largely associated with the weak intermolecular hydrogen bonding formed by MBA and aniline molecules.

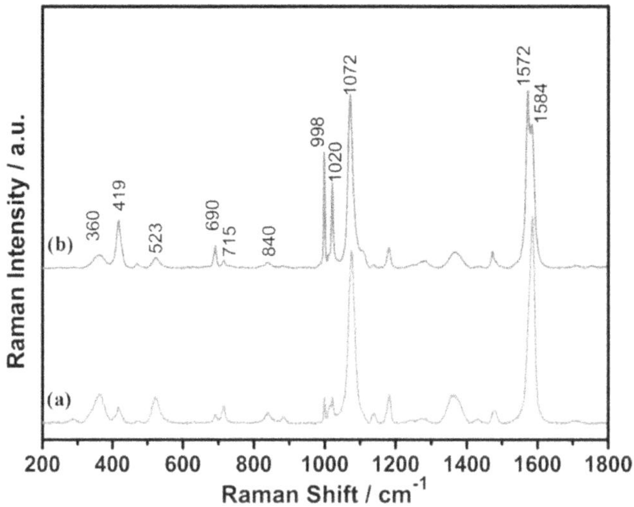

Figure 2. SERS spectra of MBA (a) in blank and (b) in the Ag/MBA/aniline system. The two SERS spectra were normalized by the band at 1075 cm^{-1}.

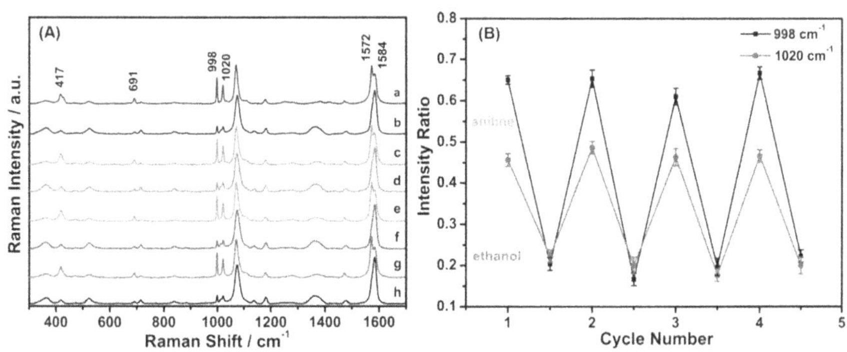

Figure 3. (A) SERS spectra of MBA dipped alternately in 10^{-3} M aniline and ethanol (from a to h). (B) Repetitive cycling of the SERS spectra for the MBA-Ag complex measured by monitoring the changes in the intensities of the bands at 998 and 1022 cm^{-1}. The SERS spectra are normalized by the band at 1075 cm^{-1}. Reprinted with permission from (Ref. (48)). Copyright (2014) American Chemical Society.

In virtue of intermolecular hydrogen bonding in the system is a non-covalent interaction and vulnerable to the changes of its surroundings, a cycled experiment was performed to verify the origin of the changes in SERS spectra of MBA. As recorded the SERS spectra of MBA in the hydrogen bonding assembled system above, a changed spectral pattern is obtained (Figure 3(A-a)). Then, rinse the Ag/MBA complex from the hydrogen bonding system to remove the aniline molecules, and after that the Ag/MBA complex was placed in the blank solution to be collected its SERS spectrum. The SERS spectrum of MBA changed as it should be under experimental conditions. Repeat the same procedure as the abovementioned for another three cycles to obtain the cycled SERS spectra of MBA shown in Figure 3(A), which indicate that it is the non-covalent intermolecular hydrogen bonding consist of MBA and aniline resulting in the repeat changes in the SERS spectrum rather than the chemical reaction between MBA and aniline molecules. In each cycle, the relative intensity of some bands in the SERS spectra changed significantly, particularly in the case of the bands at 998 and 1022 cm^{-1}, whose intensities cycled repeatedly Figure 3(B). Since the changes in the vibrational modes of MBA associated with the bands are known to reflect the interfacial contribution to the modification in the polarizability of the phenyl ring (*49*, *51*), it is reasonable that intermolecular H-bonding plays a critical role in the redistribution of charge in the MBA molecule.

Concentration-Dependent SERS Spectra of Hydrogen Bonding Assembled System

To explore the effect of hydrogen bonding in system, a concentration-dependent SERS experiment was carried out by varying the concentration of aniline. As shown in Figure 4A, the most prominent changes in the SERS spectra is the changes in relative intensities of the bands at 417cm^{-1}, assigned to the ν(C–S) mode, at 691 and 713 cm^{-1}, respectively attributed to the C–H out-of-plane deformation mode and the γ(CCC) out-of-plane bending mode, at 998 and 1022 cm^{-1}, and at 1572 and 1584 cm^{-1}, respectively. An obvious increasing trend for the relative intensities of the above-mentioned bands versus the concentration of aniline is illustrated in Figure 4B. The changes in the intensity of these in-plane and out-of-plane modes of the phenyl ring in MBA molecule can be considered a demonstration that the spectral changes are mainly determined by the modification in the electronic structure of the MBA molecule. As the hydrogen bonding formed between MBA and aniline influence the degree of conjugation, which further modifies the polarizability of the bonds for the vibration of phenyl ring in MBA, it is reasonable that the SERS intensities of the corresponding bands are related to the change of the concentration of hydrogen bonding ligand in system. Therefore, SERS spectroscopy was proved to be a powerful technique to study intermolecular hydrogen bonding with great spectral response, even the hydrogen bonding ligand in system is at trace level of concentration, whose effect is difficult to be investigated with the usage of other spectroscopy methods.

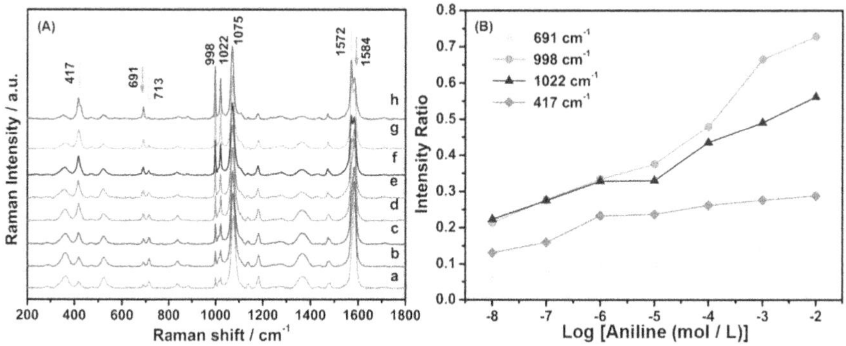

Figure 4. (A) Normalized SERS spectra of MBA by the band at 1075 cm^{-1} upon exposure to varying concentration of aniline; the concentrations are 0, 10^{-8}, 10^{-7}, 10^{-6}, 10^{-5}, 10^{-4}, 10^{-3}, and 10^{-2} M (a–h, respectively). (B) Intensities of the bands at 417, 691, 998, and 1022 cm^{-1} with respect to that of the band at 1075 cm^{-1} in the SERS spectra of (A) versus the log concentration of aniline. Reprinted with permission from (Ref. (48)). Copyright (2014) American Chemical Society.

Frequency Shifts in the SERS Spectra by Hydrogen Bonding

As is known that the formation of hydrogen bonding could change the electronic density of molecules in system, the polarizability of the bonds in the molecules would be affected consequently, which causes some noticeable changes not only in the intensities of one band but also in its frequency. From the enlarged range of the concentration-dependent SERS spectra of MBA (Figure 5), an evident shift can be observed for both the bands at 1075 and 1365 cm^{-1} with the increase of concentration of aniline. In the case of the band at 1075 cm^{-1} (Figure 5(A)), which is assigned to the in-plane ring breathing coupled with (C–S) modes, it is usually a spectral marker for monitoring modifications in the electronic structure of the phenyl-Ag complex caused by external environment factors (*51*, *52*). The downshift of this band is a verification of the effect of intermolecular hydrogen bonding on the electronic structure of the molecules. Another possible interpretation for this red shift phenomenon is the stress caused by intermolecular interaction with the capture of aniline molecules into MBA molecules in the system. The other frequency shift arises from the COO– stretching mode of MBA (Figure 4B), which exhibits a distinct red shift with increasing aniline concentrations, accompanied by the reduction in the intensity of the band simultaneously. The reduction of the intensity is mainly by the reason of the inhibition of the COO– stretching mode after forming the hydrogen bonding. Thus, the changes of this vibrational mode is a more intuitive and obvious reflection of hydrogen bonding formed between the carboxyl group of MBA and the amine group of aniline. Consequently, the changes could be certainly observed both in the frequency and the intensity of this vibration of MBA under the influence of hydrogen bonding.

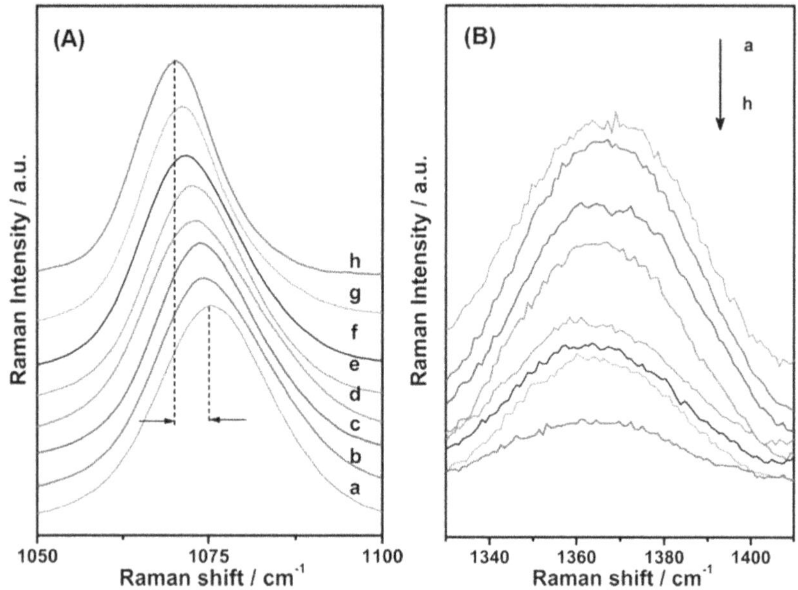

Figure 5. Details of the 1050−1100 cm^{-1} (A) and 1330−1410 cm^{-1} (B) spectral regions of the SERS spectra of MBA with different concentrations of aniline normalized by the band at 1075 cm^{-1}; the concentrations are 0, 10^{-8}, 10^{-7}, 10^{-6}, 10^{-5}, 10^{-4}, 10^{-3}, and 10^{-2} M (from a to h respectively). Reprinted with permission from (Ref. (48)). Copyright (2014) American Chemical Society.

Exploring Effect of Intermolecular Hydrogen Bonding on Charge-Transfer in System

As demonstrated in the preceding part, the relative intensity of some vibrational mode of the molecules changed associated with intermolecular hydrogen bonding, in which includes the selectively enhanced non-totally symmetry vibration (b_2 mode). The changes in the b_2 mode can be empirically considered the occurrence of a CT process in system (*45*, *46*, *53*). In this part, *p*-aminobenzenethiol (PATP), a classical probe molecule for the study of CM enhancement mechanism in SERS (*54*, *55*), was chosen to explore the effect of hydrogen bonding on CT in system (*56*). The hydrogen bonding assembled system consisted of PATP and benzoic acid (BA) molecules (Figure 6), with the same preparation procedure as it in Figure 1.

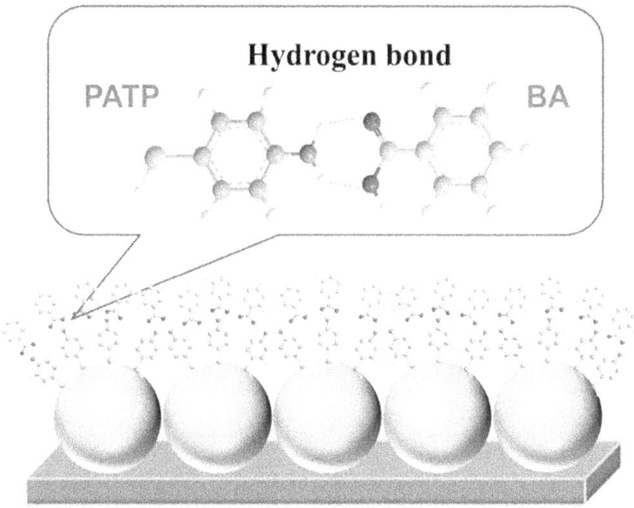

Figure 6. Schematic diagram representing the formation of a Ag–PATP–BA system.

Selective Enhancement of the b_2 Mode with Forming Hydrogen Bonding

For a SERS spectrum of PATP, the predominant bands at around 1141, 1390, and 1435 cm^{-1}, attributed to b_2-type mode, are completely different from its normal Raman spectra (*54*, *57*). The relative intensity of these bands with respect to the band at 1076 cm^{-1}, assigned to the totally symmetry vibration band (a_1-type mode) largely depends on SERS substrate as well as the external environment. In the case of the hydrogen bonding assembled system, the intensity of the b_2-type modes of PATP are dramatically enhanced with the increase of the concentration of BA molecules, as shown in Figure 7(A). This is in virtue of the formation of intermolecular hydrogen bonding between PATP and BA, leading to the modifications in the molecular electronic structure and the further changes in the polarization of related bonds of PATP, as interpreted above. Nevertheless, the discussion here will focus on the effect of hydrogen bonding on CT in the assembled system rather than the impact on other aspects.

It is acceptable that the selective enhancement of b_2 modes in the SERS spectra of molecules with C_{2v} symmetry can be empirically used as a propensity rule to recognize the participation of a CT process contributed by Herzberg-Teller effect (*37*, *46*). An equation was proposed by Lombardi *et al.* to quantitatively estimate the CT contribution to the SERS intensity, which is shown as following (*43*, *46*):

$$\rho_{\text{CT}}(k) = \frac{I^k(\text{CT}) - I^k(\text{SPR})}{I^k(\text{CT}) + I^0(\text{SPR})}$$

The index "k" in the equation represents individual lines in the Raman spectra and it may be either a totally symmetric or non-totally symmetric. Two reference lines in a spectral region were chosen for better understanding: one is totally symmetric line with the SERS signal only contributions from SPR, whose intensity is denoted I_0 (SPR), and the other is non-totally symmetric line, the intensity of which derived from CT resonance as an additional contribution to the SERS intensity excluding the contribution of SPR and is denoted I_k (CT). For a totally symmetric line, I_k (SPR) = I_0 (SPR), while for a non-totally symmetric line, I_k (SPR) is usually quite small or zero, and the SERS intensity is primarily generated from CT contributions. The more detailed description about the equation can be found in the literature.

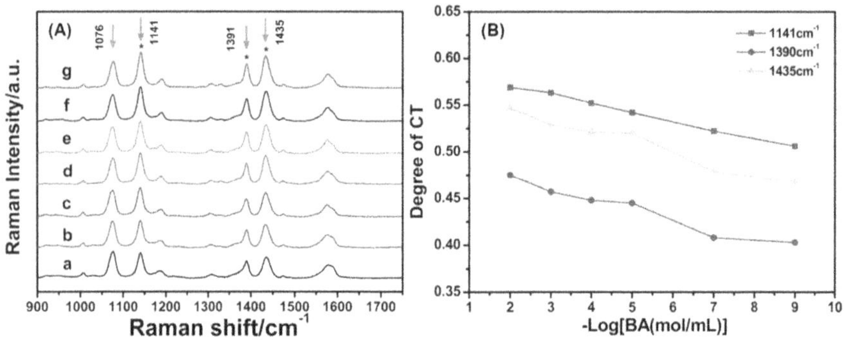

Figure 7. (A) SERS spectrum of (a) PATP molecules adsorbed on the Ag NPs and PATP molecules in the Ag/PATP/BA assembled system with a BA concentration of (b) 10^{-9}, (c) 10^{-7}, (d) 10^{-5}, (e) 10^{-4}, (f) 10^{-3}, and (g) 10^{-2} M. All spectra were measured with 633 nm excitation and normalized by the band at 1076 cm^{-1}. (B) Degree of charge transfer (p_{CT}) for the b_2 bands PATP of versus negative log of the concentration of BA, including the bands at 1141 (squares), 1390 (dots), and 1435 cm^{-1} (triangles). Reproduced from (Ref. (56)) with permission from the PCCP Owner Societies. Copyright (2014) Royal Society of Chemistry.

The laser wavelength used can excite an electron transition from the Fermi level to the molecular orbital level of PATP under experimental conditions. CT contribution plays a key role in the overall intensity of Raman signal, the CT contribution to the totally symmetric band is quite small nonetheless. According to the equation, the intensity of the a_1 band at 1076 cm^{-1} can be defined as I_0 (SPR), and the intensities of the b_2 bands at 1141, 1390, and 1435 cm^{-1} are defined as I_k (CT). Then, the degrees of CT for PATP in the Ag/PATP/BA assembled system of those b_2 bands as a function of the negative log of the BA concentration can be calculated, and plotted in Figure 7(B). It is found that the degree of CT (p_{CT}) value increase with the concentration of BA, which is caused by the hydrogen bonding between the carboxyl groups of BA and amine groups of PATP. Since the hydrogen bonding affects the electronic structure of PATP and increases the conjugation in the assembled system, the energy separation between the Fermi level of the Ag and the lowest unoccupied molecular orbital (LUMO) level may reduce. As a result, this may promote the CT transition process between the Ag NPs and the PATP

(*56*). With the concentration of hydrogen bonding ligand increased, the amount of hydrogen bonds increase, resulting in more matched CT-states involved in the CT resonance. Therefore, the enhanced intensities of the b_2-type modes of PATP and the p_{CT} value of the system are concentration-dependent.

Temperature-Dependent SERS Spectra and Its Two-Dimensional Correlation Analysis

As is known, hydrogen bonding is greatly influenced by temperature and it would be weaken and even break as rising the temperature, so does the intermolecular hydrogen bonding in the assembled system (*1*, *58*). To further verify the effect of hydrogen bonding on CT in assembled system, a temperature-dependent SERS measurement was performed in a temperature range of 25–75 °C with the increments of 5 °C, and the p_{CT} values were plotted as a function of temperature (see Figure 8(A)). In a control experiment, the blank solution was used to replace BA, and the corresponding p_{CT} values were plotted in Figure 8(B). By comparing the p_{CT} values in the two system, an evident conclusion can be draw that the p_{CT} values decrease in the hydrogen bonding assembled system with the increase of the temperature, while there was no obvious rising or downward trends for the p_{CT} values in the control experiment. These significant decrease of the p_{CT} values in the assembled system was a demonstration that the intermolecular hydrogen bonding between PATP and BA molecules facilitates the CT transition from the Ag NPs to the absorbates. When the temperature is raised, the hydrogen bonding break and the promotion to the CT transition in the assembled system reduce, resulting in the decrease in the p_{CT} value. This, in turn, indicates that the value of p_{CT} can be used to quantitatively evaluate hydrogen bonding in the system.

Two-dimensional (2D) correlation analysis, a powerful and versatile spectral analysis methods for investigating perturbation-induced variations in dynamic data (*59*, *60*), was applied to the temperature-dependent SERS spectra to elucidate the effect of intermolecular hydrogen bonding on CT process in the assembled system. The synchronous map (Figure 8(C)) shows five dominating positive auto-correlation peaks of PATP, which includes the a_1 mode at 1076 cm^{-1}, and some characteristic b_2 modes at the bands of 1141, 1390, 1435 cm^{-1}. The other peaks beyond the diagonal of the synchronous map are the correlation peaks, and the positive ones suggest that changes in SERS spectra of the PATP molecules are synchronous with the increase of the temperature, such as the ones centred at (1076, 1141), (1076, 1390), (1076, 1435). However, their corresponding correlation peaks on the asynchronous map are negative (Figure 8(D)). According to the Noda's rules (*60*), it can be concluded that changes in the b_2 bands are delayed relative to changes in the a_1 band with increasing the temperature by combining the signs of the corresponding correlation peaks in synchronous and asynchronous maps. That is, the b_2 bands are more susceptible to the effects of hydrogen bonding than the a_1 bands. Hence, it indicates that the CT transition process in the assembled system is largely affected by the intermolecular hydrogen bonding.

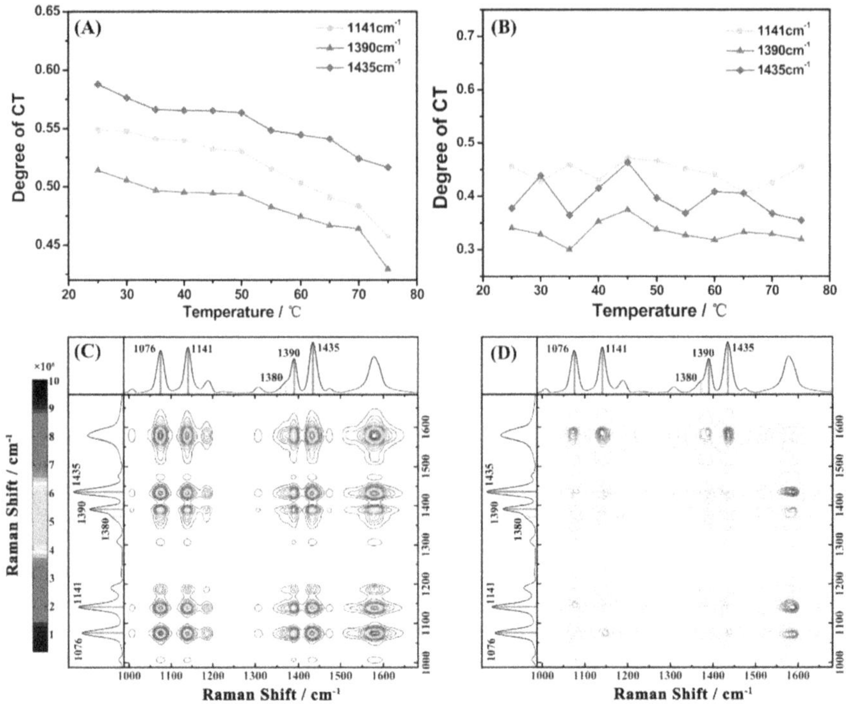

Figure 8. The degree of charge transfer (p_{CT}) for PATP molecules in the (A) hydrogen bonding assembled system and (B) blank system versus temperature via the intensities of characteristic b_2 bands at 1141 (dots), 1390 (triangles), and 1435 cm^{-1} (squares). (C) Synchronous and (D) asynchronous 2D correlation maps of the SERS spectra of PATP in the hydrogen bonding assembled system conducted at 25–75 °C with the increments of 5 °C. Adapted from (Ref. (56)) with permission from the PCCP Owner Societies. Copyright (2014) Royal Society of Chemistry.

A Label-Free Enantioselective Discrimination of Alcohols by Hydrogen Bonding

SERS has been proved to be a powerful technique for the study of intermolecular hydrogen bonding, which can reflect the subtle changes in the molecular polarizability and the CT transitions in system because of the effect of hydrogen bonding with sensitive response. On the basis of this, a label-free method for enantioselective discrimination of alcohols was proposed by using SERS spectroscopy, in which intermolecular hydrogen bonding in an assembled system plays a key role in the discrimination process (*61*). Notably, neither any chiral reagents nor the circularly polarized light (chiral light) were prerequisite in this discrimination approach.

The Discrimination of Chiral Alcohols in Hydrogen Bonding Assembled System

The hydrogen bonding assembled system for chiral discrimination was constituted by the chiral alcohols and the complexes composed by Ag NPs and achiral selector molecules, *p*-mercaptopyridine (MPY). For this chiral discrimination system, an intriguing phenomenon was found that the Ag/MPY complex displayed different SERS spectral patterns with different chiral enantiomers of 1,1,1-trifluoro-2-propanol (TFIP) (for details, see Ref. (*61*)). As shown in Figure 9(A), the SERS spectral profile of the Ag/MPY complex re-shaped distinctly with the interaction of the racemic TFIP, and the relative intensities of three pairs of bands were inversed with respect to the case interacting with the R-type enantiomer of TFIP. These inversed changes occurred in the pairs of the bands at 1009 /1096 cm^{-1}, 1202 /1220 cm^{-1}, and 1578 /1612 cm^{-1}, which include the X-sensitive mode at 1096 cm^{-1} assigned to the ring-breathing coupled with the υ(C-S) mode, the vibration related to the nitrogen in the aromatic rings at 1202 cm^{-1} ascribed to β(CH) ($9a_1$)/δ(NH) modes, and the stretching vibration υ(CC) at 1578 and 1612 cm^{-1} (*61–63*). For the change in the relative intensity of these vibrations, it is a manifestation of the modification of the molecular electronic structure by the reason of the intermolecular hydrogen bonding formed between the nitrogen atom of MPY and the hydroxyl group of TFIP (*64*).

By comparing the SERS spectra of Ag/MPY complex in different chiral environments, it is discovered that the dramatic change in the SERS profile merely occurred when the Ag/MPY complex interacted with the raceme of TFIP. While it interacted with the R-TFIP molecules, the SERS pattern is the same with the measured profiles of the complex placed in other achiral fluorine-containing or fluorine-free alcohols (Figure 9(B)). Considering the two types of enantiomers were included in the racemic TFIP, the only difference between the raceme and the R-TFIP is the S type enantiomer. These dramatic changes in the relative intensities in the SERS spectrum of MPY originate from the difference in the hydrogen bonding of the two enantiomers and MPY, either the stereoscopic configuration or the composition of the hydrogen bonding between the two enantiomeric environments. That is, the S enantiomer of TFIP, forming a differentiate configuration of intermolecular hydrogen bonds interacted with MPY in comparison to the R enantiomer in the assembled system, is more likely to be selectively distinguished. In the discrimination process, intermolecular hydrogen bonding plays a critical role, and neither any chiral reagents nor the circularly polarized light (chiral light) were involved in the assembled system. Therefore, it is reasonable that a phenylthiol molecule without a terminal nitrogen atom cannot be utilized for the discrimination of the two enantiomers of TFIP (Figure 9(C)).

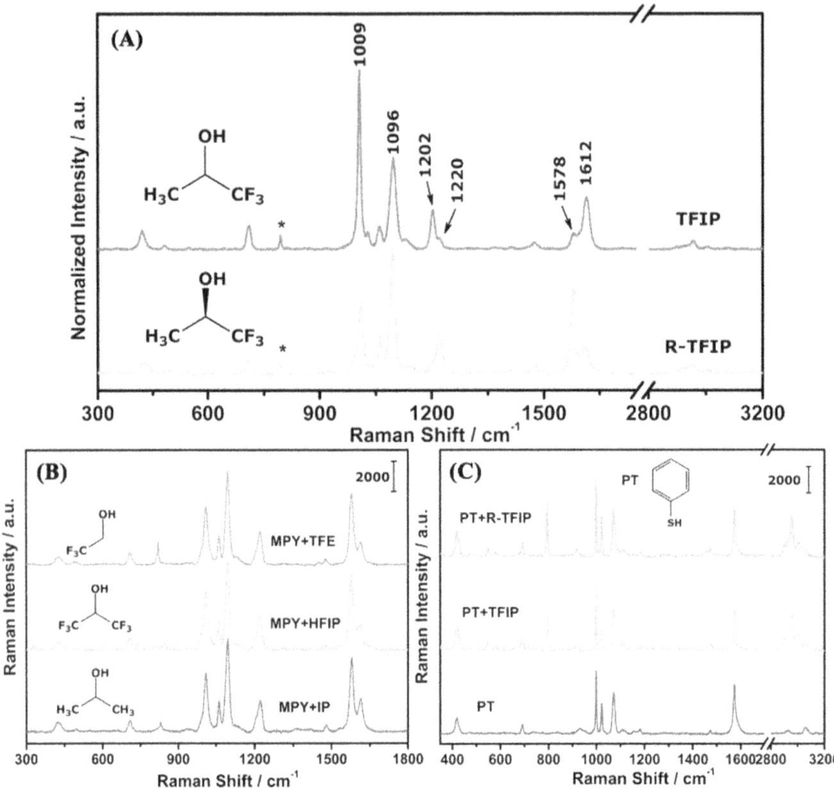

Figure 9. (A) SERS spectra of the Ag/MPY complex upon exposure to racemic TFIP and (R)-TFIP, respectively. The SERS spectra was normalized by the intensity of the band at 795 cm-1, which was assigned to TFIP (marked with an asterisk in the figure). (B) SERS spectra of Ag/MPY complex in various achiral fluoric or fluoride-free alcohols. From bottom to top; IP, HFIP, and TFE. (C) SERS spectra of PT-Ag complex (bottom), and the PT-Ag complex immersed in TFIP (middle) and R-TFIP (top) measured under the same experiment conditions as those for the MPY-Ag complex. Reproduced from (Ref (61)). Copyright (2014, John. Wiley and Sons).

Evaluating Efficacy and Universality of the Enantioselective Discrimination

An enantiomeric purity-dependent SERS experiments was conducted to examine the efficacy of the enantioselective discrimination, in which different enantiomeric purity of the R-TFIP with various ee values was studied (Figure 10(A)). It was observed that the relative intensities of some bands in the SERS spectra of Ag/MPY depend on the ee value of the TFIP mixture. The pair of bands at 1202 and 1220 cm^{-1} was chosen as an indicator, as the peak at 1202 cm^{-1} can be regarded as a direct response to the enantioselectivity through the formation of hydrogen bonding between the nitrogen atom MPY and the hydroxyl group of

TFIP. Evidently, plotting the difference in the ratio of the relative intensities of the peaks at 1202/1220 cm^{-1} against the ee value (in %) revealed a linear correlation with a coefficient of R^2= 0.9948 (see Figure 10(B)). It can be concluded that this achiral assembled system is capable of discriminating the enantiomers of a chiral alcohol by using SERS spectroscopy. Furthermore, the hydrogen bonding based label-free method is demonstrated to estimate the enantiomeric purity quantificationally as well.

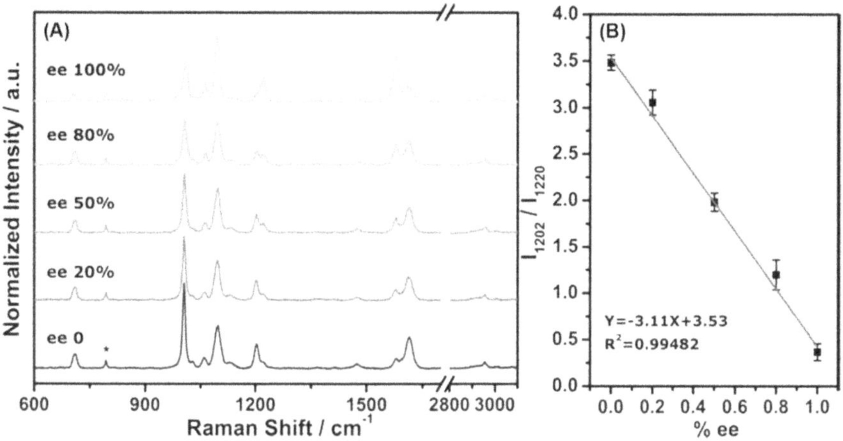

Figure 10. (A) SERS spectra of the Ag/MPY complex in the presence of TFIP with various ee values. (B) Correlation between the differences in the relative intensity ratio of I_{1202}/I_{1220} and the ee values (in %). All of the Raman peaks were normalized to the intensity of the band at 795 cm^{-1}, which was assigned to TFIP (marked with an asterisk in the figure). Reproduced from (Ref (61)). Copyright (2014, John. Wiley and Sons).

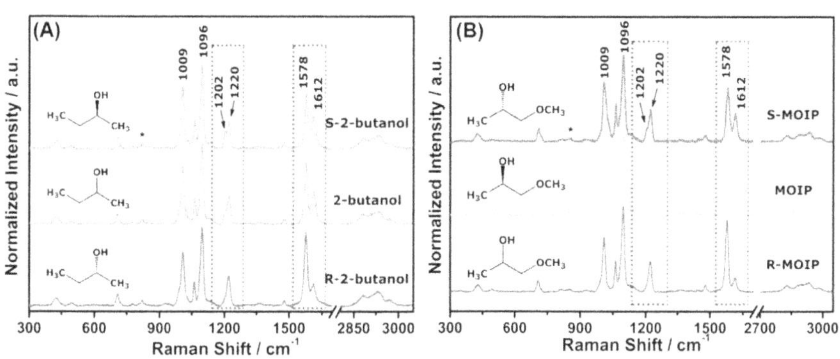

Figure 11. Normalized SERS spectra of the Ag/MPY complex interacting separately with 2-butanol (A) and MOIP (B) in their optically pure and racemic forms. Reproduced from (Ref (61)). Copyright (2014, John. Wiley and Sons).

Now that this enantioselective discrimination derives from the effect of hydrogen bonding, it should have no specific selectivity of some kind of alcohol for discrimination. The optically pure and racemic forms of two other chiral alcohols, 2-butanol and 1-methoxy-2-propanol (MOIP), were chosen to verify the universality of this label-free discrimination method for various kinds of alcohols, as depicted in Figure 11. It can be investigated that the same changing tendency appearing in the aforementioned bands in the SERS spectra of the Ag/MPY complex. A distinct decrease in the intensity of the band at 1578 cm^{-1} and a simultaneous increase in the intensity of the bands at 1612 cm^{-1} were observed, when the Ag/MPY complex interacted with the S enantiomers, or even in the racemic mixture containing S enantiomers. Moreover, the enantioselective discrimination indicator, i.e., the ratio of the intensity of the bands at 1202 and 1220 cm^{-1}, in both 2-butanol and MOIP increased with the enantiomeric purity of S enantiomer. However, the magnitude of the changes in these bands greatly diminished compared with the case with TFIP. As fluorine is the most electronegative element, the trifluoromethyl group in a TFIP molecule has strongly electron-withdrawing properties. Thus, the hydroxyl group in TFIP has a greater tendency to form hydrogen bonds than those of 2-butanol and MOIP, which contain the less electronegative ethyl (CH_3CH_2) and methoxy (CH_3O) groups, respectively.

Preliminary Exploration of the Possible Mechanism

As discussed, the origin of the discrimination behavior for chiral alcohols by SERS spectra should be associated with the protonation of the terminal nitrogen atom in MPY molecule when interacted with different enantiomers through hydrogen bonding. In the process of the formation of hydrogen bonding, an difference in orientation or composition between the two assembled systems constructed by Ag/MPY complexes and either of the enantiomers of a chiral alcohol, resulting in a distinct difference in the SERS spectra. From the SERS spectra obtained in Figure 9 and 11, it is noted that the a_1 modes of MPY, such as the ring-breathing mode at 1009 cm^{-1}, the ring-breathing/υ(CS) mode at 1096 cm^{-1}, the β(CH)/δ(NH) mode at 1202 cm^{-1}, and the υ(CC) mode at 1612 cm^{-1}, were enhanced in a certain extent with the interaction of S enantiomer, whereas the b_2 mode at 1578 cm^{-1}, attributed to υ(CC), is tremendously diminished in comparison with the spectra measured under the conditions with the existence of R enantiomers. It is thus considered that the subtle difference between the two assembled systems causes different energy states that may induce differentiated CT processes. The CT contributions in the system amply the very slight differences between the two assembled system owing to the chiral environment, and realized the enantioselective discrimination by the remarkable differences in the relative intensities of SERS spectra.

According to the CT mechanism, it can be inferred that the S-type enantioselective discrimination process in assembled system increased the contribution of the Frank–Condon term and simultaneously inhibited the Herzberg–Teller term (*44*, *46*, *53*, *55*). Nevertheless, the R enantiomer, in spite of hydrogen bonding, may have a difference in either orientation or composition compared to the case in S enantiomers, leading to some suitable CT state being involved in the CT transition. SERS enhancement of the MPY molecules may still be greatly influenced by the Herzberg–Teller effect. Thus, these enantioselective phenomenon in SERS spectra is dominated by the CT enhancement mechanism, based on the effect of intermolecular hydrogen bonding in the system. Further explanations for this label-free method for enantioselective discrimination continue to be required. Despite of this, it is believed that this study opens a new avenue leading to the development of novel enantiosensing strategies.

Summary

In this chapter, SERS spectroscopy was demonstrated as an advantageous technique for investigating the effect of hydrogen bonding on the molecular electronic structures and the CT in system. It was found that the changes in the vibration frequency and the relative intensities occurred in the SERS spectra, duo to the modifications of molecular polarizability by hydrogen bonding. Meanwhile, the CT process generated between the substrates and the adsorbates was significantly facilitated by hydrogen bonding. By taking advantage of the influence of hydrogen bonding, a label-free method for enantioselective discrimination by SERS spectroscopy was proposed, in which the relative intensities of the molecules in system largely depended on the chiral environment. The preliminary explanation of the intriguing enantioselective discrimination has been made that different CT transition may be involved for different enantiomeric environment. Nevertheless, more work will be required to deeply understand the mechanism and to fully exploit the application of this enantioselective discrimination method.

Acknowledgments

The authors thank their coworkers in the studies reported in this article for their contributions, and the great help from Prof. Xiaoxia Han and Dr. Zhi Yu for instruction and suggestion during writing of this chapter. The related research was supported by the National Natural Science Foundation (21273091, 21221063, and 21327803) of China, the Specialized Research Fund for the Doctoral Program of Higher Education (20110061110017), the 111 project (B06009), and the Development Program of the Science and Technology of Jilin Province (20110338, 20130305005GX).

References

1. Desiraju, G. R.; Steiner, T. In *The Weak Hydrogen Bond in Structural Chemistry and Biology*; Oxford University Press, Inc.: New York, 1999.
2. Prins, L. J.; Reinhoudt, D. N.; Timmerman, P. *Angew. Chem., Int. Ed.* **2001**, *40*, 2382–2426.
3. Kovacs, A.; Szabo, A.; Hargittai, I. *Acc. Chem. Res.* **2002**, *35*, 887–894.

4. Nagy, P. I. *Int. J. Mol. Sci.* **2014**, *15*, 19562–19633.
5. Mayoral, M. J.; Bilbao, N.; Gonzalez-Rodriguez, D. *ChemistryOpen.* **2016**, *5*, 10–32.
6. Sherrington, D. C.; Taskinen, K. A. *Chem. Soc. Rev.* **2001**, *30*, 83–93.
7. Schuster, P.; Mikenda, W. In *Hydrogen Bond Research*; Springer-Verlag: New York, 1999; Vol. 130, pp 1−103.
8. Reedijk, J. *Chem. Soc. Rev.* **2013**, *42*, 1776–1783.
9. Buck, U.; Huisken, F. *Chem. Rev.* **2000**, *100*, 3863–3890.
10. Maeda, H.; Wang, Y.; Ozaki, Y.; Suzuki, M.; Czarnecki, M. A.; Iwahashi, M. *Chemometr. Intell. Lab.* **1999**, *45*, 121–130.
11. Singh, D. K.; Mishra, S.; Ojha, A. K.; Srivastava, S. K.; Schlucker, S.; Asthana, B. P.; Popp, J.; Singh, R. K. *J. Raman Spectrosc.* **2011**, *42*, 667–675.
12. Takahashi, M. *Crystals* **2014**, *4*, 74–103.
13. Lange, K. M.; Aziz, E. F. *Chem. Asian J.* **2013**, *8*, 318–327.
14. Ninet, S.; Datchi, F.; Klotz, S.; Hamel, G.; Loveday, J. S.; Nelmes, R. J. *Phys. Rev. B* **2009**, *79*, 100101.
15. Srinivasan, R.; Feenstra, J. S.; Park, S. T.; Xu, S.; Zewail, A. H. *J. Am. Chem. Soc.* **2004**, *126*, 2266–2267.
16. Kapur, G. S.; Cabrita, E. J.; Berger, S. *Tetrahedron Lett.* **2000**, *41*, 7181–7185.
17. Piotrowicz, E. K.; Dziewulska, K.; Stangret, J. *J. Phys. Chem. B* **2010**, *114*, 5810–5818.
18. Kolesov, B. A.; Mikhailenkobc, M. A.; Boldyreva, E. V. *Phys. Chem. Chem. Phys.* **2011**, *13*, 14243–14253.
19. Auer, B. M.; Skinner, J. L. *J. Chem. Phys.* **2008**, *128*, 224511.
20. Chiashi, S.; Murakami, Y.; Miyauchi, Y.; Maruyama, S. *Jpn. J. Appl. Phys.* **2008**, *47*, 2010–2015.
21. Hearne, G. R.; Zhao, J.; Dawe, A. M.; Pischedda, V.; Maaza, M.; Nieuwoudt, M. K.; Kibasomba, P.; Nemraoui, O.; Comins, J. D.; Witcomb, M. J. *Phys. Rev. B* **2004**, *70*, 2806–2810.
22. Nie, S.; Emory, S. R. *Science* **1997**, *275*, 1102–1106.
23. Chen, H. Y.; Lin, M. H.; Wang, C. Y.; Chang, Y. M.; Gwo, S. *J. Am. Chem. Soc.* **2015**, *137*, 13698–13705.
24. Zrimsek, A. B.; Wong, N. L.; Van Duyne, R. P. *J. Phys. Chem. C* **2016**, *120*, 5133–5142.
25. Lalanne, P.; Andanson, J. M.; Soetens, J. C.; Tassaing, T.; Danten, Y.; Besnard, M. *J. Phys. Chem. A* **2004**, *108*, 3902–3909.
26. Wilson, S. O.; Tran, N. T.; Franz, A. K. *Organometallics* **2012**, *31*, 6715–6718.
27. Abraham, M. H.; Abraham, R. J.; Byrne, J.; Griffiths, L. *J. Org. Chem.* **2006**, *71*, 3389–3394.
28. Chang, X.; Zhang, Y.; Weng, X.; Su, P.; Wu, W.; Mo, Y. *J. Phys. Chem. A* **2016**, *120*, 2749–2756.
29. Joseph, J.; Jemmis, E. D. *J. Am. Chem. Soc.* **2007**, *129*, 4620–4632.
30. Heaton, R. J.; Madden, P. A. *Mol. Phys.* **2008**, *106*, 1703–1719.
31. Fang, Y.; Wu, G.; Wang, P. *Chem. Phys.* **2012**, *393*, 140–147.

32. Mah, P. T.; Fraser, S. J.; Reish, M. E.; Rades, T.; Gordon, K. C.; Strachan, C. J. *Vib. Spectrosc.* **2015**, *77*, 10–16.
33. Kneipp, K.; Moskovits, M.; Kneipp, H. *Surface-Enhanced Raman Scattering: Physics and Applications*; Springer: Berlin, 2006.
34. Ozaki, Y.; Kneipp, K.; Aroca, R. In *Frontiers of Surface-Enhanced Raman Scattering: Single Nanoparticles and Single Cells*; Wiley: Hoboken, NJ, 2014.
35. Camden, J. P.; Dieringer, J. A.; Zhao, J.; Van Duyne, R. P. *Acc. Chem. Res.* **2008**, *41*, 1653–1661.
36. Moskovits, M. *Rev. Mod. Phys.* **1985**, *57*, 783–826.
37. Lombardi, J. R.; Birke, R. L. *Acc. Chem. Res.* **2009**, *42*, 734–742.
38. Campion, A.; Kambhampati, P. *Chem. Soc. Rev.* **1998**, *27*, 241–250.
39. Ikeda, K.; Suzuki, S.; Uosaki, K. *J. Am. Chem. Soc.* **2013**, *135*, 17387–17392.
40. Yang, L.; Jiang, X.; Ruan, W.; Zhao, B.; Xu, W.; Lombardi, J. R. *J. Phys. Chem. C* **2008**, *112*, 20095–20098.
41. Stiles, P. L.; Dieringer, J. A.; Shah, N. C.; Van Duyne, R. R. *Annu. Rev. Anal. Chem.* **2008**, *1*, 601–626.
42. Jensen, L.; Aikens, C. M.; Schatz, G. C. *Chem. Soc. Rev.* **2008**, *37*, 1061–1073.
43. Lombardi, J. R.; Birke, R. L.; Lu, T. H.; Xu, J. *J. Chem. Phys.* **1986**, *84*, 4174–4180.
44. Albrecht, A. C. *J. Chem. Phys.* **1961**, *34*, 1476–1484.
45. Yamamoto, Y.; Itoh, T. *J. Raman Spectrosc.* **2016**, *47*, 78–88.
46. Lombardi, J. R.; Birke, R. L. *J. Phys. Chem. C* **2008**, *112*, 5605–5617.
47. Lee, P. C.; Meisel, D. *J. Phys. Chem.* **1982**, *86*, 3391–3395.
48. Wang, Y.; Ji, W.; Sui, H.; Kitahama, Y.; Ruan, W.; Ozaki, Y.; Zhao, B. *J. Phys. Chem. C* **2014**, *118*, 10191–10197.
49. Guerrini, L.; Pazos, E.; Penas, C.; Vázquez, M. E.; Mascareñas, J. L.; Alvarez-Puebla, R. A. *J. Am. Chem. Soc.* **2013**, *135*, 10314–10317.
50. Chen, T.; Wang, H.; Chen, G.; Wang, Y.; Feng, Y.; Teo, W. S.; Wu, T.; Chen, H. *ACS Nano* **2010**, *6*, 3087–3094.
51. Saikin, S. K.; Olivares-Amaya, R.; Rappoport, D.; Stopac, M.; Aspuru-Guzik, A. *Phys. Chem. Chem. Phys.* **2009**, *11*, 9401–9411.
52. Zayak, A. T.; Hu, Y. S.; Choo, H.; Bokor, J.; Cabrini, S.; Schuck, P. J.; Neaton, J. B. *Phys. Rev. Lett.* **2011**, *106*, 083003.
53. Centeno, S. P.; López-Tocón, I.; Roman-Perez, J.; Arenas, J. F.; Soto, J.; Otero, J. C. *J. Phys. Chem. C.* **2012**, *116*, 23639–23645.
54. Park, W. H.; Kim, Z. H. *Nano Lett.* **2010**, *10*, 4040–4048.
55. Zhou, Q.; Li, X.; Fan, Q.; Zhang, X.; Zheng, J. *Angew. Chem., Int. Ed.* **2006**, *45*, 3970–3973.
56. Wang, Y.; Ji, W.; Yu, Z.; Li, R.; Wang, X.; Song, W.; Ruan, W.; Zhao, B.; Ozaki, Y. *Phys. Chem. Chem. Phys.* **2014**, *16*, 3153–3161.
57. Osawa, M.; Matsuda, N.; Yoshii, K.; Uchida, I. *J. Phys. Chem.* **1994**, *98*, 12702–12707.

58. Otto, K. E.; Hesse, S.; Wassermann, T. N.; Rice, C. A.; Suhm, M. A.; Stafforst, T.; Diederichsen, U. *Phys. Chem. Chem. Phys.* **2011**, *13*, 14119–14130.
59. Noda, I.; Ozaki, Y. In *Two-Dimensional Correlation Spectroscopy-Applications in Vibrational and Optical Spectroscopy*; John Wiley and Sons: Chichester, U.K., 2004.
60. Noda, I. In *In Handbook of Vibrational Spectroscopy*; Chalmers, J., Griffiths, P., Eds.; John Wiley & Sons: New York, 2002; Vol. 3.
61. Wang, Y.; Yu, Z.; Ji, W.; Tanaka, Y.; Sui, H.; Zhao, B.; Ozaki, Y. *Angew. Chem., Int. Ed.* **2014**, *53*, 13866–13870.
62. Hu, J.; Zhao, B.; Xu, W.; Li, B.; Fan, Y. *Spectrochim. Acta A* **2002**, *58*, 2827–2834.
63. Chao, Y.; Zhou, Q.; Li, Y.; Yan, Y.; Wu, Y.; Zheng, J. *J. Phys. Chem. C* **2007**, *111*, 16990–16995.
64. Fioroni, M.; Burger, K.; Roccatano, D. *J. Chem. Phys.* **2003**, *119*, 7289–7296.

Chapter 7

Advances in Biofunctional SERS-Active Nanoparticles for Future Clinical Diagnostics and Therapeutics

Steven Asiala,† Lee Barrett,†
Samuel Mabbott,† and Duncan Graham*

Centre for Molecular Nanometrology, WestCHEM, Pure and Applied Chemistry, University of Strathclyde, Technology and Innovation Centre, 99 George Street, Glasgow, United Kingdom, G1 1RD
†These authors contributed equally to the manuscript.
***E-mail: duncan.graham@strath.ac.uk**

The synergy afforded by the combination of biofunctionalised nanoparticles and surface enhanced Raman scattering (SERS) has expanded the analytical toolbox for clinical diagnostics and therapeutics. Since their inception, SERS-active nanoparticles have been developed into biofunctional nanoparticles (BFNPs) using a variety of methods to attach biomolecules and pacification layers to nanoparticles to enable detection of various diseases or cancers *in vitro* and *in vivo*. However, while there are many reports of the use of BFNPs for diagnostic or therapeutic applications, very few are implemented in a "real" clinical setting, for example, detection of disease biomarkers in tissue or the delivery of drugs to affected cells. This review covers recent advances made in the development of BFNPs for SERS-based detection of clinical samples using *in vitro* and *in vivo* methods.

Introduction

In the ongoing fight against diseases, the need for novel, rapid tools for diagnosis to improve treatment is omnipresent. Much attention is being paid to developing methods and technologies aimed at providing clear and accurate information to clinicians, with the larger goal of improving patient outcomes through stratified medicine.

One area of research that has been mined for its potential to aid in clinical settings is vibrational spectroscopy (*1*). Vibrational spectroscopies have been historically used to identify functional groups present in a given molecule due to their "fingerprint" spectra, but more recently to detect and identify biomarkers linked to a number of diseased states. The most commonly used forms of vibrational spectroscopy are infrared (IR) (*2*, *3*) and Raman (*4–6*), with each applied to a wide range of biomedical applications (*1*, *7*). Furthermore, nonlinear vibrational spectroscopies that utilize multi-photon, pump-probe arrangements, such as coherent anti-Stokes Raman (CARS) (*8*) and stimulated Raman (SRS) (*9*, *10*) have been developed with an eye toward biomedical imaging and clinical application (*11*, *12*).

Concurrent with advances in vibrational spectroscopy methods has been the advancement of nanomaterials for biomedical applications, including imaging, photothermal therapy (PT), drug delivery, and sensing (*13*). Nanomaterials, such as metallic nanoparticles, quantum dots, and carbon-based materials have physical and chemical properties that differ from the bulk and can be exploited for a specific biomedical or clinical task. More specifically, plasmonic metal nanoparticles have been of great service in the realms of vibrational spectroscopy and imaging due to their ability to drastically increase the sensitivity of Raman spectroscopy. Surface-enhanced Raman Spectroscopy (SERS) (*14*) has brought Raman into the realm of ultrasensitive techniques, with reports of single-molecule detection (*15*, *16*). The prospects of using SERS for detection in clinical applications (*17–19*), including cancer detection and imaging (*20*) have been the subject of recent reviews (*21*).

In addition to increased Raman sensitivity, another important asset of nanomaterials is the ability to tailor their surface chemistry (*22*). Nanoparticles can be functionalised with beacons, or reporter molecules (*23–25*), for tracking and detection, encapsulated for anti-fouling, stabilization and pacification in biological environments (*26*, *27*), and targeted for specific biomarkers with the addition of antibodies (*28*), DNA (*29*), peptides (*30*), and aptamers (*31*). When each of these four elements is present— a plasmonic metal nanoparticle, reporter, pacification layer, and biomolecule—a biofunctional nanoprobe (BFNP) is created as depicted in Figure 1. This strict definition of a BFNP will be used as a guide for the work reviewed herein.

The goal of this review is to highlight the most recent progress made toward the utilization of BFNPs for vibrational spectroscopy in clinical applications. This will include a brief review of especially relevant plasmonic nanoparticle properties and functionalization strategies, followed by a survey of the use of BFNPs for *in vitro* and *in vivo* diagnostics and therapeutics.

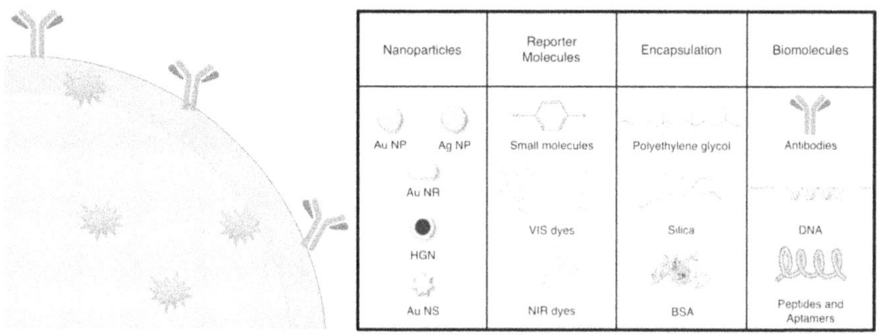

Figure 1. A cartoon representation of the criteria for biofunctional nanoparticles (BFNP) for clinical applications. BFNPs consist of a plasmonic metal nanoparticle, a reporter molecule, an encapsulation or pacification layer, and a biomolecule for specific targeting. Shown are common particles, reporters, encapsulation materials, and biomolecules. Bovine serum albumin structure from PDB reference 4F5S (32).

Plasmonic Nanoparticles

Nanoparticles bridge the gap between bulk material and molecular structure (*33*) offering unique properties which make them appealing to the biosciences, especially in the area of diagnostics and therapeutics. Nanoparticles are often defined as having dimensions ranging from 1 to 100 nm. Particles synthesised on this scale have found use across both physical and biological sciences due to properties such as increased surface area and the ability to exchange energy from light to heat.

However, one phenomenon that makes plasmonic nanoparticles appealing to physicists and spectroscopists alike is the confinement of electrons on their surface. The sinusoidal oscillation of electrons on the surface of the single metallic particles is termed a localised surface plasmon. To take advantage of increased field effects, nanoparticles have to be subjected to a resonant frequency. UV-Vis spectrophotometry is one analytical technique that can identify λ_{max} of the particles at which both absorption and scattering is at its maximum; this frequency is often termed the local surface plasmon resonance (LSPR). It is the LSPR that is responsible for the enhancement seen in surface enhanced Raman scattering (SERS) when metallic nanoparticles are used.

Many reviews have been written on SERS (*34*, *35*) and its initial observation (*36*), therefore it is beyond the scope of this review to fully explain the intricacies of the technique. Instead, only a brief discussion of the mechanisms will be given with further sections, focusing on the use of nanoparticles to enable the method for *in vitro* and *in vivo* biological studies.

Electromagnetic Effect

As mentioned previously, the tuning of light to the LSPR of nanoparticles causes an increase in electric field intensity as a consequence of surface plasmon excitation. This increase in the electric field is extended perpendicularly to the particle surface, causing it to interact with any molecules in close proximity. The molecules become bathed in a freely moving electron cloud, causing them to become polarized. It is the interaction of these electrons with the molecule that gives rise to intense Raman scattering effects. This mechanism contributes an enhancement factor of around 10^6, although this is estimated to be much less for a single particle than for a multi-nanoparticle system. Estimations of the increased electric field can be quantified using generalised Mie theory (GMT), and, although mathematically complex, the theory is explained exceptionally well by Stockman *et al.* (*37*)

Chemical Effect

Whilst electromagnetic enhancement is accountable for the greatest contribution to SERS enhancement, it alone cannot justify the high level of enhancement seen in many systems. The chemical mechanism of enhancement occurs when a physical bond is formed between the nanoparticle and the analyte. This effect is often interpreted as a HOMO and LUMO interaction by which charge can be transferred between the particles and analyte.

SERS can also be facilitated on a range of two- and three-dimensional nanostructured metallic substrates (*38–40*), but the mobility of nanoparticles is paramount to their *in vivo* and *in vitro* application. Therefore, only nanoparticles will be discussed in this review and in particular gold and silver nanoparticles as these are the most commonly used in SERS.

General Nanoparticles Synthesis

Gold nanoparticles have been used throughout history most commonly to adorn decorative objects such as stained glass. It was not until 1857 that the first scientific evaluation of colloidal gold was made by Michael Faraday. Synthesis of colloidal gold was achieved by reduction of ‘a moderately strong solution of chloride of gold’ using phosphorus (*41*). It was the development of the colloidal suspension that led to the discovery of the Faraday-Tyndall effect, whereby longer-wavelength light is transmitted whilst shorter wavelengths become more reflected via scattering. Most interesting, however, is that Faraday’s particles have remained stable for over 150 years, whilst modern researchers can often only maintain optical stability for a number of months.

Two primary techniques can be used for the synthesis of gold and silver nanoparticles. Physical methods, which use processes such as thermal decomposition and laser ablation have been shown (*42–49*), but more often chemical methods are employed, where a chemical species is used to reduce metallic salts. It is a common theme that strong reducing agents produce small nanoparticles whilst nanoparticles reduced using weaker reagents are bigger.

Examples of reducing agents include $NaBH_4$ (*50, 51*), hydroquinone (*52, 53*), ascorbic acid (*54, 55*) and, of course, the most commonly used citrate, which will be discussed in greater depth in future sections. In a move to incorporate the principles of green chemistry, it has also become increasingly popular to use natural products in the synthesis of nanoparticles, utilizing extracts from a variety of plant leaves, roots and tissues as reducing agents (*56–58*).

Once synthesised, the maintenance of colloidal stability is paramount to ensure the longevity of the particles. Reagents used for stabilising the nanoparticles can be placed into two categories; electrostatic or steric. In electrostatic stabilisation, a charged species is often used to interact with the nanoparticle, whilst simultaneously repelling other nanoparticles. When steric stabilisation is carried out, bulky chemical species such as polyethylene glycol (PEG) (*59–62*), bovine serum albumin (*63*), or chitosan polysaccharide (*64–67*), are used but many more species are also available (*27*). It should be noted that some of the chemicals mentioned such as citrate and chitosan act as both a reducing and stabilising reagent.

Nanoparticle: Elemental Composition

Nanoparticles can and have been synthesised using a variety of metals, but it is important to remember that for SERS to occur, the LSPR of the metals must be excited across the UV-Vis-NIR region. A few papers report the use of Pt, Pd, Ru, Rh and transition metals such as Ni for SERS substrates, but many of these are non-nanoparticle based approaches, using mostly electrodes as the active interface (*68–79*). Copper nanoparticles (CuNPs) are, however, well mentioned in the literature. Whilst chemical reduction is the general route for nanoparticles synthesis, Muniz-Miranda *et al.* showed that CuNPs with a broad size range (majority 3-9 nm) and a LSPR band around 588 nm could be synthesised via laser ablation using a 1064 nm laser (*80*). SERS enhancement of both Phen and Bipy were achieved using by coupling the CuNPs to lasers with 532 nm and 785 nm excitations. Silicon-Hydrogen bond assisted assemblies of copper nanoparticles have also been proven capable of enhancements of 2.29×10^7 and a RSD <20% using Rhodamine 6G as an analyte (*81*). It is possible to form a variety of different copper nanoparticle shapes as demonstrated by Li *et al.,* whose nanocube and nanoplate structures termed CuTe (copper telluride) yielded respectable enhancement factors of 1.5×10^6 and 10^6 when using hydrophobic Nile Red as the reporter (*82*). Only a handful of papers have shown that SERS-active copper nanoparticles can be stabilised with biocompatible capping agents. The synthesis of gelatin stabilised spherical nanoparticles (20 nm) was demonstrated by Zhang *et al.* Chemical reduction of a Cu (II) salt was carried out using hydrazine, and the resultant gelatin capped particles were shown to have a LSPR of ~572 nm. SERS analysis of crystal violet gave a calculated enhancement factor of 1.02×10^3 when excited using a 514.5 nm laser (*83*). SERS enhancement has also been demonstrated using chitosan stabilised copper nanoparticles again with dimensions of ~20 nm, here the G and D bands of the biomolecule at 1382 cm^{-1} and 1851 cm^{-1} were shown to have much greater intensity on the copper compared to using silver (*84*).

Whilst there is certainly development on-going in the use of a variety of metals to form nanoparticles that elicit a SERS response, very little has been contributed to the application of the nanoparticles towards biological platforms. This is still an area which is very much dominated by gold and silver nanoparticles (*85*, *86*) and hence the main focus of this review.

In order for a nanoparticle to function within a biomedical capacity it must have properties that make them desirable to the application. Initially the nanoparticles must be biocompatible, this means that dependent on which route they are administered into the body they must not be excreted before reaching their intended site of action. The particles cannot be cytotoxic to healthy cells and must be non-reactive. These are assets that will all prevent unforeseen damage to a patient. The surface of the particles should also be amenable to modification, allowing the attachment of a range of targeting biomolecules, stabilising agents and Raman reporters. Finally, it is essential that the LSPR is 'tunable'; for particles of a given chemical composition, tuning of the LSPR can be achieved by changing their shape, morphology and/or size. Tunability of the nanoparticles into the NIR (near-IR) area of the electromagnetic spectrum is hugely important in studies where the particles are contained within blood or tissue phantoms as NIR light can penetrate much more deeply and efficiently than visible radiation (*87*, *88*).

Due to their vast usage and compatibility from hereon, only silver and gold nanoparticles synthesised into the shape of spheres, rods, shells and hollow structures will be discussed as this review is not designed to cover all the various shapes and compositions of nanoparticles used in the field.

Nanoparticle Shapes: Nanospheres

Spherical nanoparticles are by far the most readily used particles and there are a diverse set of wet-methods that can be employed for their synthesis. However, citrate reduced and stabilised nanoparticles more often than not, form the base for *in vivo* and *in vitro* applications. The reduction method developed by Turkevich in 1951 and later refined by Frens in 1972 utilises citrate as both the reducing and capping reagent (*89–91*). In order to obtain gold particles of various sizes and greater reproducibility, various groups have adjusted a number of variables involved in the citrate reduction. Li *et al.* established an environmentally friendly method for the production of gold citrate nanoparticles by increasing the pH of the reaction solution with NaOH and lowering the temperature. The group were able to slow the reaction rate sufficiently by changing both pH and temperature simultaneously to reduce nanoparticle polydispersity (*92*). Similar methods were also used by Bastus *et al.* for the inhibition of secondary nucleation using a seeded growth route; the method developed is capable of synthesising gold nanospheres ranging from 10 to 200 nm. More significantly, the increase in size relates to a red shift in the extinction spectrum, suggesting the large spheres could be used *in vivo* applications (*93*). Adjustments to the original Turkevich protocol carried out by Schulz *et al.* boast a narrow particle size distribution of 5-8% standard deviation and reproducible diameters with led than 3% variation. To achieve monodispersity and reproducibility, a citrate buffer and 0.02 M EDTA were employed with the latter improving the circularity of the nanoparticles (*94*). One of the major issues with synthesising nanospheres in this manner is the limitation to aqueous conditions. This was overcome by Brust *et al.* when they demonstrated the two phase (water-toluene) reduction of gold chloride in the presence of a phase-transfer agent (tetraoctylammonium bromide) by sodium borohydride in dodecanethiol (*95*).

The synthesis of silver nanospheres can also be carried out using citrate reduction and stabilization (Figure 2). This was first demonstrated by Lee and Meisel in the early 1980s (*96*). The citrate capped particles can also be synthesised in a range of sizes from ~15 nm to 190 nm with plasmon excitation spanning from 401-715 nm (*97*). An alternative method of silver nanosphere synthesis that is also readily employed is the hydroxylamine/sodium hydroxide reduction developed by Leopald and Lendl (*98*). Both methods, citrate and hydroxylamine reduction, are ideal for preparation of biofunctional nanoparticles because both chemical moieties demonstrate weak interaction with the metallic surface. Spherical nanoparticles often form the basis from which anisotropic particles are synthesised, with the more complex structures demonstrating a greater shift in LSPR.

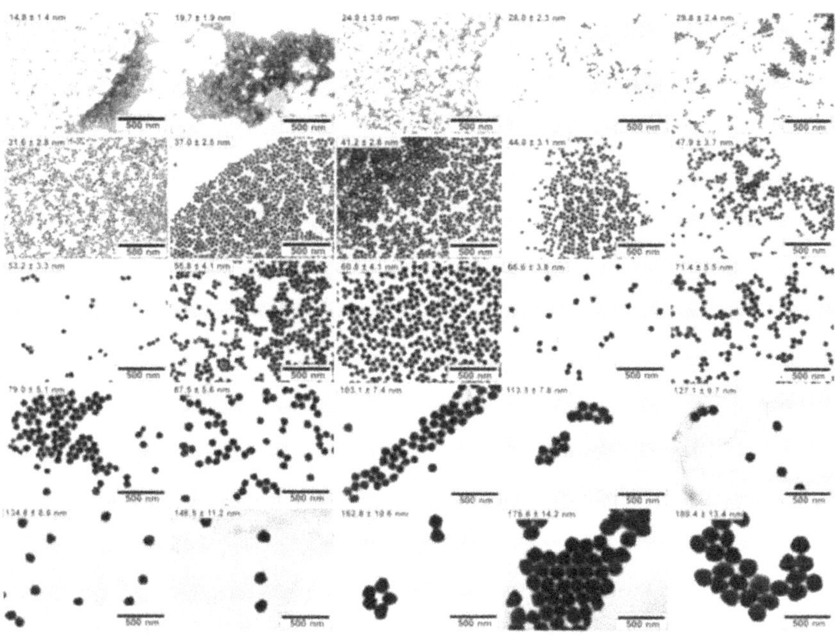

Figure 2. Colloidal solutions of Ag NPs of differing sizes. Shown is seeded-growth synthesis of silver nanoparticles. Transmission electron microscopy images of the seed particles and those obtained after different growth steps. The particle size increases from 14.8 ± 1.4 nm to 189.4 ± 13.4 nm, and its concentration decreases from ~ 2.3 x 10^{12} to ~ 1.1 x 10^{10} NP/mL. Adapted with permission from ref. (97). Copyright 2014 ACS Publications.

Nanoparticle Shapes: Rods

Rod shaped nanoparticles present an attractive prospect for facilitating SERS as they have plasmon bands that reside in both transverse and longitudinal directions. These bands can be tuned across the Vis-NIR range by varying the aspect ratio of the particles. The ability to shift the plasmon bands makes the rods useful in *in vivo* studies and makes them extremely flexible with regards to tailoring to a specific excitation frequency. Orendorff, used a variety of gold and silver nanorods with differing aspect ratios to demonstrate the importance of the overlap between the plasmon and the laser excitation in 2006. Rods whose plasmon matched the excitation source gave rise to $10\text{-}10^2$ greater Raman enhancement (*99*). The importance of the plasmonic overlap has been reiterated several times (*100*, *101*). The most common method of nanorod synthesis is seed-mediated growth. Murphy together with co workers explored wet chemistry methods for the production of gold nanorods (*102*). In the seed-mediated approach, ascorbic acid, a weak reducing agent, is used to reduce chloroauric acid in the presence of silver nitrate, CTAB and nanospherical seeds (~3 nm) to produce rod shaped nanostructures. This method was later improved upon by El-Sayed (*103*). It was discovered that by reducing the number of nanospherical seeds greater aspect ratios could be achieved, but to further improve the yield of rods, it was essential to switch from citrate capped seeds to CTAB capped. Zhoa and co-workers demonstrated that increasing the amount of ascorbic acid used during the synthesis of the rods decreased aspect ratios and that long term storage was achieved in the presence of CTAB (*104*) In an effort to simplify the seed-mediated method, seedless methods have been developed using the hydrophobic species paradioxybenzene. Rods synthesised using this method can exhibit plasmon excitations up to 1400 nm (*105*).

The seed-mediated and seedless approaches can also be applied to the synthesis of silver nanorods and wires, where the aspect ratio is often on the micron scale (*106*, *107*). It has also been shown that it is possible to control the reaction rate and aspect ratio of silver nanorods by subjecting them to a red shifted excitation wavelength from the LSPR of spherical seed particles (*108*). One issue that arises when using either gold or silver nanorods as a functional particle is the presence of CTAB at its surface. The surfactant interacts with the rods much more strongly than other capping agents, so the chemistry employed to functionalise the particle needs to be much more considered and often more intricate. Fortunately methods are being developed which do not involve CTAB. For example Lin *et al.* demonstrated that potassium tartrate can be used to reduce silver nitrate in the presence of poly vinyl pyrrolidone to produce nanorods via a seedless methodology (*109*). The review by Zubarev and co workers excellently outlines functionalisation and synthesis strategies for application-based gold nanorods (*110*).

Nanoparticle Shapes: Nanoshells and Hollow Structures

Recently, much research has been focused on the development of nanoshells as these structures, like rods, are also highly tunable whilst consisting of a more favourable spherical shape. Traditionally, nanoshells consist of a dielectric core coated with either a gold or silver layer. Both the core size and the shell thickness can be adjusted in order to achieve the desired optical properties. The core is often synthesised from silica using the Stöber method. The resultant particles are highly monodispersed and spherical, providing an ideal template for a layer noble metal deposition at the surface. To grow the metallic layer, small nanoparticle seeds are attached to the core that has been functionalised using an amino or mercaptosilane. The core size can be varied between 50 and 500 nm, whilst adjusting the concentration controls the thickness of the metallic layer between 5 to 30 nm. Using this method it is possible to shift the plasmon of the particles across the visible into the mid-IR region.

Hollow structures have similar optical properties to the nanoshells except that they are synthesised using a sacrificial core. The first nanoshells were synthesised by *Sun et al.*, via galvanic displacement of gold onto a silver core. The plating of one metal onto another relies on the electrode potential of the core being lower than that of the shell material. The sacrificial cobalt core has been synthesised under inert conditions to afford particles, which are homogenous, spherical, and of high quality. Hollow gold particles synthesised in this way have an LSPR tunable from the visible to NIR region (550 – 820 nm). Tuning of the hollow gold LSPR can be achieved by varying the core diameter and/or changing the concentration of either the gold or capping agent. The bandwidth of the LSPR is dependent on the uniformity of the structures composite parts, poor uniformity often results in broad spectral distribution. Increasing core size and reducing the gold shell thickness promotes a red shift in absorbance. More recently, Bedics *et al.* demonstrated that hollow gold with an LSPR residing at 720 nm could be modified with chalocogenopyrylium dyes whose absorbance resides in the NIR region to realise picomolar detection when subjected to a 1280 nm laser (*23*). This demonstrates the determination for researchers to extend towards the lower frequency end of the electromagnetic spectrum (*24*).

SERS-Active BFNPs for Clinical Diagnostics

Combining NPs with biomolecules creates molecularly specific probes that have promising applications in the biomedical fields, particularly in molecular diagnostics and medical imaging (*108*, *111*). In these fields, the ability to control the size and shape of NPs, and therefore the optical properties, provides a unique advantage for tailoring NPs to a specific application. Controlling the optical properties of NPs also improves their functionality as potential SERS probes which can provide greater sensitivity and multiplexing capability for the analysis of clinically relevant targets, such as cancers and infectious diseases (*20*, *112*).

In the following sections, research on biofunctional nanoparticles (BFNPs) towards clinical diagnostics and therapeutics is discussed. This focus of this discussion is primarily on research that has demonstrated the analytical validity of the BFNPs in immunoassays and have expanded the sample pool to include patient samples in order to evaluate the methodology in a clinical setting. Due to the breadth of literature available in this regard, recent developments (~ 5 years) in the field of SERS-based clinical diagnostics using BFNPs are discussed in two sections: *in vitro* and *in vivo* applications.

In Vitro Diagnostic (IVD) Applications

This section aims to provide a detailed overview of the recent immunoassays which have incorporated SERS-active BFNPs, with particular emphasis on the methods used to functionalise the NPs and how the BFNPs perform with clinical samples.

Extrinsic Raman Labels and Antibody-Functionalised NPs

The use of BFNPs as detection probes in SERS-based immunoassays has been the subject of much research in recent years. Porter and co-workers developed a sandwich immunoassay for the detection of pancreatic cancer (PC) biomarkers in human serum by utilizing extrinsic Raman labels (ERLs) (*113*). In this case, the ERLs consisted of 60 nm AuNPs functionalised with a mixed monolayer of Raman-active reporter molecules and antibodies specific to mucin protein, MUC4, which were immobilised by using a bifuntional linker, dithiobis (succinimidyl propionate) (DSP). This linker contains a thiol group for attachment to the AuNP surface and a succinimidyl ester that readily reacts with the the lysine ($-NH_3^+$) residues on the antibody to form a stable amide bond. The ERLs were then used to detect MUC4 levels in the sera of PC patients using SERS (Figure 3), which has been previously unattainable by conventional methods, such as enzyme-linked immunosorbent assay (ELISA) and radioimmunoassay (RIA). Recently, Krasnoslobodtsev *et al.* improved upon the detection of MUC4 using ERLs by incorporating an overcoat layer of PDMS in the sandwich immunoassay, which drastically improved the SERS response obtained from serum samples (*114*).

Porter and co-workers also developed a multiplexed sandwich immunoassay array panel for the detection of pancreatic adenocarcinoma (PA) using ERLs prepared using the same method described previously (*115*). Two different Raman reporters were chosen to correspond to two different biomarkers for PA – MMP-7 and CA 19-9. The authors reported that using SERS for the analysis of patient samples generated lower LODs compared to a conventional ELISA, further establishing the usefulness of SERS as a potential clinical diagnostic tool.

Dufek *et al.* reported an analogous immunoassay approach to those mentioned previously. However, here, the authors employ NeutrAvidin-functionalised 70 nm NPs and SERS for the competitive detection of a vitamin D metabolite, calcitriol, which is a marker for renal failure in patients exhibiting low serum concentrations (*116*). Here, a preliminary assessment of the SERS performance was evaluated using patient samples containing low and high concentrations of calcitriol. The results indicated that the SERS-based detection platform compared favorably to the more conventional RIA detection. However, the authors note that further studies with a wider set of patient samples are required in order to validate their SERS platform for routine clinical diagnostic use.

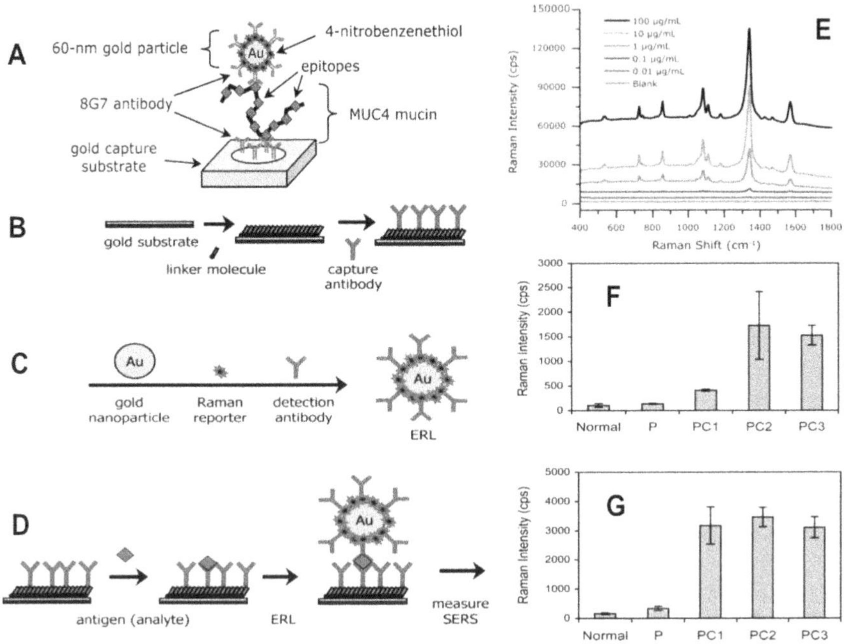

Figure 3. SERS-based immunoassay readout using BFNPs. (A) Schematic representation of the sandwich immunoassay approach for the detection of MUC4 mucin. (B) Preparation of the capture substrate to extract and concentrate target antigen in solution. (C) Preparation of extrinsic Raman labels (ERLs). (D) Steps involved in the preparation of the immunoassay with SERS readout. (E) SERS spectra acquired at different cell lysate concentrations containing MUC4. (F) SERS detection of MUC4 and (G) CA19-9 in pooled sera from normal individuals and patients with pancreatitis (P) or pancreatic cancer (PC1, PC2 and PC3). Adapted with permission from ref. (113). Copyright 2011 American Chemical Society.

Li *et al.* have devised a method using a 3D-plasmonic architecture resembling a Au triangle nanoarray chip that captures the target antigen, followed by detection by SERS in a sandwich immunoassay format using antibody-functionalised NPs (*117*). The NPs consisted of Au nanospheres or Au nanostars that were functionalised with a mixed monolayer of malachite green isothiocyanate (MGITC) as the Raman reporter molecule and 3-triethoxylsilylpropyl succinic anhydride (TEPSA) in order to introduce –COOH groups to the NP surface. The antibody was then coupled to the –COOH groups via EDC/ NHS coupling chemistry. The antibody-functionalised nanospheres/ nanostars were used to detect IgG as a proof of concept showing that the nanostars resulted in the best SERS signals compared to the nanospheres. The Au nanostars were then used to detect vascular endothelial growth factor (VEGF) in clinical blood plasma samples. The SERS results obtained were compared to those collected from a commercially available ELISA kit for VEGF, which showed similar results between the two methods. However, in this case, while the results were similar, the SERS immunoassay has several advantages over the ELISA method, such as shorter detection times, fewer wash steps and more robust readout from complex sample matrices.

Chen *et al.* have reported a sandwich immunoassay using BFNPs for the detection of carcinoembryonic antigen (CEA) in the serum of 26 colorectal cancer patients (*118*). The authors used Au/Ag core-shell NPs functionalized with 4-mercaptobenzoic acid (4-MBA) and anti-CEA antibodies. The antibodies were immobilized onto the NP surface at elevated pH in borate buffer, followed by blocking the NP surface with BSA. The BFNPs were then used to detect CEA in serum samples using SERS and electrochemical luminescence for comparison. It was found that SERS was able to detect lower concentrations of CEA (LOD = 5 pg/ mL), suggesting this SERS sandwich immunoassay is a viable tool for clinical detection. Chen *et al.* also utilized a similar BFNP method for the detection of latent-membrane protein 1 (LMP1) in nasopharyngeal tissue sections from 34 cancer patients (*119*). This study also demonstrates the promising potential for SERS to be used in clinical analysis for LMP1 over the more conventional immunohistochemistry staining methods.

Wu *et al.* achieved the simultaneous detection of two cancer marker, p53 and p21, in blood serum using SERS and Au@Ag core-shell nanorods (Au@Ag NRs) (*120*). The nanorods were functionalised with two different Raman reporters, 4-MBA and 5,50-dithiobis (2-nitrobenzoic acid) (DTNB), and poly (allylamine hydrochloride) (PAH) in order to introduce amine functionality to the NR surface. The two different Au@AgNRs were then functionalized with Abs corresponding to the two biomarkers using glutaraldehyde crosslinking. Serum samples were then immobilised onto functionalised glass slides and the SERS-active Ab-Au@AgNRs added to detect the antigen in serum. The reported LOD was less than 1 pg/mL, which is lower than the average level of the biomarkers in healthy human plasma, thereby demonstrating the clinical applicability of this method.

Magnetic Separation with BFNPs

Immunoassay methods can have various different formats. The methods previously discussed in this review article have consisted of sandwich immunoassays using substrates coated with a capture antibody. However, another promising format for clinical IVDs is magnetic separation. Chon *et al.* used hollow Au nanospheres (HGNs) and magnetic beads to simultaneously detect two lung cancer biomarkers in blood serum (*121*). Au nanospheres were functionalised with MGITC and X-rhodamine-5-(and-6)-isothiocyanate (XRITC) in order to render them SERS-active. The antibodies were attached via EDC/ NHS coupling chemistry to dihydrolipoic acid (DHLA), which was added to the SERS-active nanospheres to introduce –COOH moieties for conjugation. The functionalised Au nanospheres were then used to evaluate the levels of CEA and α-fetoprotein (AFP) in serum samples from 5 different patients using SERS.

The levels of each biomarker present in the clinical samples , as determined using the SERS method, were in good agreement with those reported using a commercially available chemiluminescent microparticle immunoassay (CMIA). The SERS immunoassay method reported by Chon *et al.* has several advantages over the standard CMIA method, namely, the simultaneous detection of multiple biomarkers, quicker assay times and smaller reagent consumption, making SERS nanoprobes based on magnetic separation a viable format towards routine clinical detection.

Chon *et al.* extended their magnetic bead-based SERS immunoassay to the detection of myocardial infarction biomarkers in human serum (*122*). MGITC and XRITC functionalised Ab-Au nanospheres were again utilized for the simultaneous detection of creatine kinase (CK-MB) and tropnonin I (cTnI). A total of 19 patient samples were screened for the presence of these biomarkers and the magnetic bead SERS-based approach was reported to be 100-1000 times more sensitive than the conventional electro-chemiluminescent assay method. The work of Chon and co-workers demonstrates the viability of SERS with magnetic separation for the detection of various different biomarkers in clinical samples.

More recently, Choo and co-workers have validated the use of SERS-based immunoassays for the early diagnosis of rheumatoid arthritis (RA) (*123*). Here, SERS nanotags were prepared using HGNs functionalized with SH-PEG-COOH and human IgG antibodies, which were covalently attached via EDC/NHS coupling. The IgG-HGN SERS nanotags formed a sandwich immunoassay with cyclic citrullinated peptide (CCP)-functionalized magnetic beads for the detection of auto-CCP – an autoantibody used as a biomarker for RA. In terms of performance, it was reported that the SERS-based immunoassay showed good agreement in the higher anti-CCP concentration range (>25 UmL^{-1}) with a commercially available ELISA kit. In the lower concentration range (<25 UmL^{-1}), the SERS-based immunoassays out-performed the ELISA kit, demonstrating the robustness of the SERS method for diagnostic tests. To this end, 74 clinical blood samples were screened using the SERS immunoassay and compared to the ELISA results. It was reported that there was good correlation between SERS and ELISA data in the anti-CCP positive group (n=31, > 25 UmL^{-1}). However, it was found that the SERS immunoassay showed greater reproducibility in the negative anti-CCP group (n=43, <25 UmL^{-1}) compared to the ELISA, which further establishes the potential of SERS as a viable alternative to ELISAs in a clinical diagnostic setting.

Zboril and co-workers have adopted a similar magnetic bead approach for the determination of human IgG in blood (*124*). Here, iron oxide (Fe_3O_4) nanocomposites were mixed with silver nitrate to coat the surface of the nanocomposites with silver ions. Streptavidin was then covalently attached to a carboxy-terminated poly (ethylene) glycol linker using EDC/ NHS coupling chemistry. Finally, a biotinylated anti-IgG antibody was added to the Fe_3O_4@Ag@streptavidin nanocomposites. In contrast to the methods reported previously, where a Raman-active reporter molecule has been typically added as part of the NP functionalisation process, the authors have opted to omit the use of a Raman reporter, instead relying on the presence of unique spectral bands in the SERS spectrum when the target antigen is present. This method was applied to the analysis of IgG in human blood samples obtained from two patients using a finger prick method. IgG levels were determined to be 9-10 g/L using this SERS method, which was in good agreement with amounts of IgG in a healthy population. An abnormal level of IgG in blood is indicated in the pathology of disease, thereby validating the promising application of FeO@Ag@Streptavidin@IgG nanocomposites in IVDs.

Aptamer-Functionalised NPs

Aptamers are an appealing alternative to antibodies for immunoassay readout due their high affinity, stability and specificity for target analytes compared to antibodies. While it is outside the scope of this review to discuss aptamers in detail, several reviews are available on the topic of aptamers and their current trends in diagnostics (*125–127*). Recently, He *et al.* reported the use of a magnetic chitosan and Ag/chitosan NP as SERS probes for the detection of platelet-derived growth factor BB (PDGF BB) in human serum samples (*128*). Aptamers were attached to the NP surface by adding chitosan (CS) to the NPs, followed by glutaraldehyde, which forms a Schiff base between the amines located on CS and the 5'-terminus of the DNA aptamer. This set-up facilitated the capture of PDGF BB in a sandwich immunoassay format analogous to that reported in the previous section using magnetic capture/separation. The analytical sensitivity using this method was shown to be 3.2 $pgmL^{-1}$. He *et al.* tested their method in human serum samples. 3 healthy and 3 cancer patient samples were analysed using SERS and an ELISA for reference. Using their SERS-based immunioassay the authors could determine that the serum PDFG levels in cancer patients were higher than that of healthy patients and the results obtained were in good agreement with the ELISA reference method. In addition, while the clinical relevance of this method has been demonstrated, the authors have also reported how their aptamer SERS-based immunoassay compares to other aptamer incorporating methods in the literature, such as colourimetric, fluorescence and chemiluminescence. Of the 9 other methods compared, the SERS-based immunoassay reported by He *et al.* showed the lowest limit of detection demonstrating the clinical feasibility of using SERS for PDGF-BB detection.

In Vitro Therapeutics

A logical progression can be made from targeting biomarkers for diagnosis to targeting biomarkers for the delivery of some therapeutic measure, such as photothermal (PT) heating (hyperthermia) or drug delivery (*129*, *130*). The effort to combine diagnostic and therapeutic capabilities simultaneously is often referred to as "theranostics (*130*, *131*)." To achieve the desired therapeutic effect, varied strategies have been employed with regard to nanoparticle shape and functionalisation. Given that a complete review of nanoparticles as theranostics is outside of the scope of this review, this section will focus exclusively on the use of BFNPs for combined, targeted imaging and therapeutic applications.

One example of the power of BFNPs in theranostics *in vitro* is in the detection of circulating tumor cells in whole human blood samples (*132*). In this study, silver-coated gold nanorods (NRs) were coated with one of four Raman reporter molecules, poly(ethyleneglycol) (PEG), and one of four antibodies for different cancer biomarkers— epithelial cell adhesion molecule (EpCAM), CD44, Keratin18, and insulin-like growth factor (IGF). The NRs were shown, by using both SERS and PT imaging, to accumulate on the surface of cultured MCF-7 breast cancer cells, as well as specifically target the MCF-7 cells in spiked blood samples. While the PT component of the Ag/Au NR is used only in an imaging role here, the ability to target cancer cells with multiplexed BNFPs featuring a PT effect is an exciting development. A similar method for detecting CTCs was recently contributed by Wu. *et al.*, who used Au particles functionalised with 4-mercaptobenzoic acid (MBA), reductive bovine serum albumin (rBSA) and folic acid (FA) to target cells in rabbit blood (*133*). By optimizing experimental conditions the limit of detection was pushed to 5 cells/mL of sample.

With regard to particle functionalisation, the work of Song *et al.* highlights a novel, biocompatible strategy (*134*). Gold NRs were functionalised with a Raman reporter, PEG, and a ring-opening initiator. Lactide was added, which formed polylactide in the presence of the initiator, leading to the formation of plasmonic vesicles consisting of multiple BFNPs. Anti-EpCAM was subsequently added to the outside of the vesicle, allowing for targeting of CAM-positive Hep38 or CAM-negative PLC/PRF/5 carcinoma cells. Further, an anti-cancer drug, doxorubicin (DOX) was included on the inside of the vesicle. The CAM-positive cells were selectively targeted as demonstrated by SERS and dark-field microscopy, and laser irradiation caused nearly 100% target cell death due to synergistic PT (hyperthermia) damage and drug (DOX) delivery.

The desire for therapeutic utility has also lead to the investigation of different BFNP shapes, including nanotriangles (*135*), nanorattles (*136*), Au-Ag core-shell spheres (*137*), and nanostars (*138*). Boca-Farcau *et al.* (*135*) used chito nanotriangles with a Raman reporter and folic acid to target folate receptors of human ovarian carcinoma (NIH:OV-CAR-3). SERS measurements with 532 nm excitation and dark-field images showed successful targeting. By changing the laser source to 800 nm, hyperthermia was successfully induced due to the triangles' near-IR LSPR. A single wavelength (785 nm) PT approach was demonstrated by Chen *et al.* (*136*) using gold nanorattles with Raman reporter/PEG/estrogen receptor alpha (ERa) functionalisation, targeting the estrogen receptor in MCF-7 cells. The rattles' LSPR can be tuned into the near-IR region by controlling shell thickness in the Au/Au core-shell structure. Extended irradiation caused a measureable drop in cell viability for cells incubated with immune-rattles vs. a control PEG-rattle. Gold-silver core-shell particles were functionalized with aptamers for lung adenocarcinoma cells (A549) by Wu and coworkers (*137*). The aptamer functionalised particles showed specific affinity for the targeted cells when compared to other cancer cell lines, and when exposed to NIR radiation, local heating caused diminished cell viability at relatively low power densities. A novel, photodynamic treatment utilizing silica- and peptide-coated nanostars was shown by Fales *et al.* (*138*) The presence of a cell-penetrating peptide (TAT) allowed for enhanced particle uptake by BT549 breast cancer cells as demonstrated by Raman imaging. To induce cell death, this work made use of a photosensitiser, protoporphyrin IX, which causes the local production of reactive oxygen species (ROS) when illuminated with an appropriately filtered mercury lamp. Taken in the context of other work in the theranostic field (*131*), these *in vitro* results using BFNPs show the incremental progress being made toward PT and drug delivery therapies in clinical applications.

In Vivo Diagnostics

In parallel with developments in *in vitro* biofunctional nanoprobe (BFNP) assays, an effort has been made to utilise the properties of BFNPs *in vivo* for diagnostic and therapeutic clinical applications. In fact, reports of *in vivo* work are often accompanied by complementary *in vitro* and *ex vivo* studies. Many of the benefits that make BFNPs incredibly useful *in vitro*, such as low cell toxicity (*139–142*) and high sensitivity detection via SERS (*14*), have been demonstrated in living systems. An evaluation of important and more recent uses of BFNPs in the diagnosis of cancer and other diseases with an eye toward the clinic follows.

A considerable amount of attention has been paid to the use of BFNPs in the diagnosis of cancers. A great impact was made by the work of Qian *et al.* in 2008 (*143*). The BFNPs in this case were Au NPs coated with an organic Raman reporter and stabilised by a PEG shell. Biofunctionality was achieved by attaching the antibody for the epidermal growth factor receptor (EGFR), which is over-expressed in head-and-neck squamous cell carcinoma tumors. In mice with xenograft tumors (Tu668), tail-injected tumor targeting BFNPs and non-targeted NPs were spectroscopically shown to have similar uptake by the liver, but drastically disparate affinity for the tumor. Raman spectra obtained from the tumor mass with 785 nm excitation clearly showed the signature of the targeted BFNP reporter molecule, while the non-target was absent. This result was confirmed by measuring particle distribution in multiple organs via inductively coupled plasma-mass spectrometry (ICP-MS). Subsequently, an *in vitro* method for targeting EGFR has also been reported using silica-coated BFNPs (*144*).

Samanta *et al.* (*145*) and Harmsen *et al.* (*146*) have both demonstrated the influence on sensitivity of the Raman reporter chosen for *in vivo* applications of BFNPs. In the former, a library of NIR-absorptive Raman reporters based on tricarbocyanine derivatives was screened to determine which derivative would perform best for SERS detection (CyNAMLA-381). This molecule was functionalised onto Au NPs and encapsulated in bovine serum albumin (BSA) for stability and glutaraldehyde for later antibody conjugation. Single-chain variable fragment (scFv) antibodies for human epidermal growth factor 2 (HER2) were utilised, as HER2 is over-expressed in the cell line studied (SKBR-3). In vivo Raman spectroscopy confirmed the specific targeting of a xenograft tumor after tail-vein injection in a mouse model. The later work focused on calcogenopyrylium-based reporters (*146*). Similarly, this work sought to find an optimised reporter based on SERS intensity. Anti-EGFR antibodies were conjugated to silica-coated particles with each the calcogen compound and a commercially available dye (IR-792), and the BFNPs intravenously injected into mouse models. While both particles targeted the tumoral tissue, signal obtained from the calcogen out-paced the standard dye by a factor of three. Continued study of Raman reporter molecules should lead to even more sensitive measurements.

In vivo, multiplexed detection using non-targeted, silica SERS NPs had been demonstrated (*147*), but the first report of an *in vivo* multiplexed detection of targeted BFNPs came in an oral squamous cell carcinoma mouse model. Maiti *et al.* expanded on previous pilot animal studies (*145*, *148*), creating biocompatible BFNPs with near-IR reporters, BSA encapsulation, and conjugated antibodies for tumor targeting (*149*). Two BNFP types were functionalised with tumor-specific anti-EGFR, and a third with anti-HER2, which is expressed in low quantities in the specific cell line. Similar to the work of Qian *et al.* (*143*), clear concentration of the EGFR-targeted probes in the tumor mass was shown via SERS measurements, contrasting with the liver, where all three tags were found.

Dinish and coworkers have more recently demonstrated an extension of this multiplex capability by targeting BFNPs for multiple cancer targets (*150*). The BFNPs in these studies were Au NPs functionalised with one of three Raman reporters; each 'flavor' was encapsulated with PEG and functionalised with an antibody for one of three breast cancer biomarkers —EGFR, CD44, and TGF beta-receptor II (TGFβRII). A notable difference in this work was the delivery method utilised; in this case, an inter-tumoral injection was made into the MDA-MB-231 breast cancer xenograph, as opposed to the more common tail vein injection, in an attempt to minimise damage to healthy organs. Raman measurements at regular intervals confirmed the presence of each of the three BFNPs in the mass, and that targeted BFNPs remained in the tumor much longer (>48 h) than controls (<6 h). The work of Sinha *et al.* also demonstrates this selective retention of EGFR-targeted particles vs. non-target particles in tumoral tissue (human glioma U251 and epimeriod A431) (*151*). The researchers measured the SERS spectrum from particles used to stain excised tissue through successive washes, using this data to determine the binding potential and retention dynamics of the various particle and tissue combinations.

Ratiometric quantification of three BFNPs was demonstrated by Wang *et al.* (*152*) Mixtures of anti-EGFR, anti-HER2, and control Ab (isotype) BFNPs were applied topically to two tumor types—A431, known to overexpress EGFR with moderate HER2, and SkBr3, with overexpressed HER2 and moderate EGFR. The tumors were implanted into mouse models, and direct classical least squares (DCLS) regression was applied to Raman data to calculate anti-EGFR/isotype, anti-HER2/isotype, and anti-EGFR/anti-HER2 probe ratios in each tumor and in surrounding tissue (Figure 4). This work demonstrated the feasibility of simultaneously monitoring multiple biomarkers *in vivo* with BFNPs.

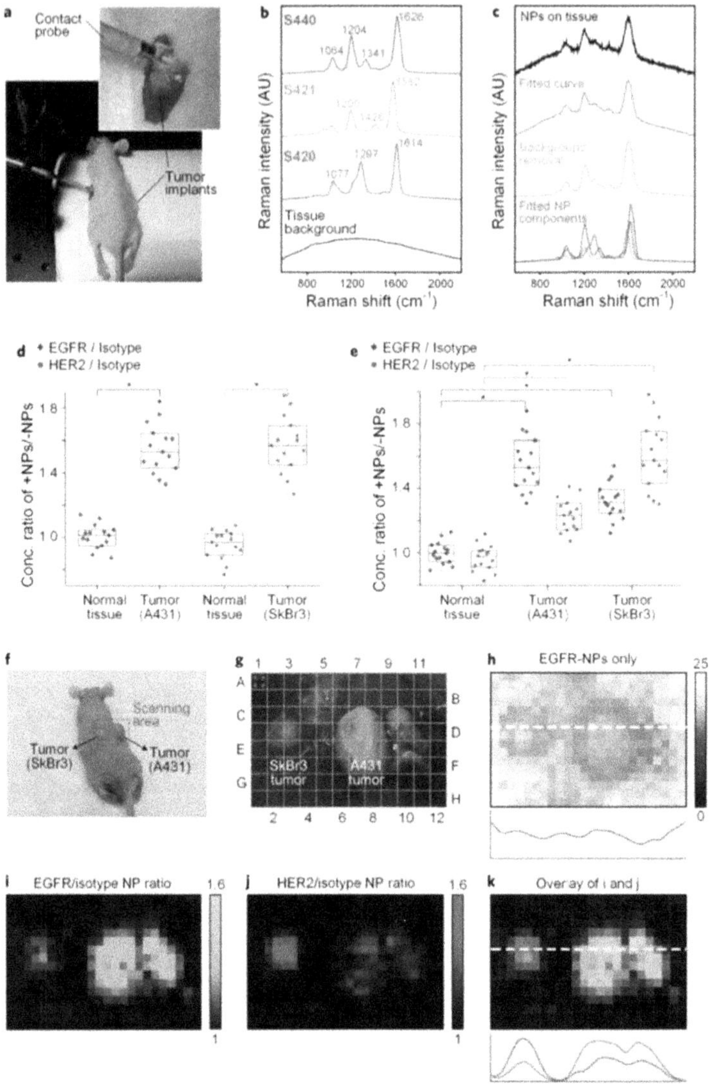

Figure 4. In vivo monitoring of multiplexed NPs applied topically to tumor implants. (a) Surgically exposed tumors and Raman probe (inset). (b) Reference BFNP spectra. (c) DCLS processing procedure for analysis of Raman spectra. Measurement of multiplexed, BFNP concentration ratios for (d) 2- and (e) 3-component mixtures is shown. (f) Mouse with two xenograft tumors. (g) Photograph of image-grid experimental scan area, with coordinates. (h) Map of absolute NP concentration (inverse contrast). (i) Map of EGFR/isotype BFNP concentration ratio. (j) Map of HER2/isotype BFNP concentration ratio. (k) Overlay of maps i and j, showing selectivity of anti-EGFR BFNPs for SkBr3 the tumor, and anti-HER2 BFNPs affinity for A431. Adapted with permission from ref. (152). Copyright 2014 World Scientific Publishing Co./Imperial College Press.

In addition to the development of non-invasive, *in vivo* tagging, there have also been efforts made to detect targeted and non-targeted probes with next-generation imaging and detection techniques. Many of these techniques aim to maximize the tissue penetration depth of the excitation photons by choosing laser wavelengths greater than 633 nm, pushing closer to the near-IR window in biological tissue. This underscores the need for BFNPs with LSPR in the near-IR. One such technique couples SERS from nanotags coupled with spatially offset Raman spectroscopy (SORS) (*153*) to perform surface-enhanced, spatially offset Raman Spectroscopy (SESORS) (*154*). Spatially offset Raman spectroscopy relies on the collection of spectra at points apart from the incident excitation, detecting diffuse photons that have been scattered by the sub-surface in multilayered samples. The combination of SERS and SORS allowed for the discrimination of non-targeted multiplexed probes 20 mm deep in tissue analogs and detection of a single probe at depths 45-50 mm. A similar study has shown probe detection through samples with varying bone thickness (*155*). Further, a fiber optic endoscopic Raman probe capable of measuring signal from non-targeted probes has been demonstrated (*156*), used in a pilot study, and further developed to include the capability to circumferentially scan (*157*). An endoscopic instrumental configuration was used by Wang *et al.* to expand on their previous work using multiplexed, ratiometric data analysis (*158*). As before (*152*), multiple flavors of targeted antibody-conjugated BFNPs (anti-EGFR, anti-HER2 and istotype) were administered to xenograft tumors A431, SkBr3, and U251), but this time with the tumors graft onto rat esophagi. BFNP SERS-signal ratios in the respective tumors were found to track well with expression of EGFR and HER2 in the tissue via cross validation techniques. Examples of small animal (*159*), widefield (*160*), and hand held probe (*161*) instruments, such as the SpectroPen of Mohs *et al.* (*162*), have also been shown, each with the goal of eventual clinical application. Though BFNPs as defined herein were not explicitly utilized in the study, the probe developed by Karabeber *et al.* (*161*) to guide brain tumor surgery through SERS spectra shows significant translational potential.

There is also interest in detecting particles using multiple imaging techniques in concert. Bardhan *et al.* used a multi-modal particle, consisting of an Au core coated in Fe_3O_4 nanoparticles, encapsulated in a fluorophore-doped (indocyanide green) silica shell and PEG, and biofunctionalised with antibody for HER2 (*163*). Differences in particle uptake by tumors over expressing and low expressing HER2 were observed through both near-IR fluorescence and magnetic resonance imaging. Kircher *et al.* have also shown an exciting *in vivo* application with non-targeted, gadolinium ion (Gd^{3+}) shell tags detected in glioblastoma bearing mice by magnetic resonance imaging (MRI) pre-surgery, photoacoustic and Raman imaging during surgery, and *ex vivo* Raman imaging of excised tissue to confirm clean margins (*164*). Photoacoustic imaging combined with SERS was also used by Dinish *et al.* to perform multimodal biosensing of BFNPs in tumor bearing mice (*165*). Similar to their previous work (*150*) and that of Sinha (*151*), the researchers also performed time dependent measurements to measure the difference in retention and update of targeted and non-targeted BFNPs in both diseased and healthy tissues.

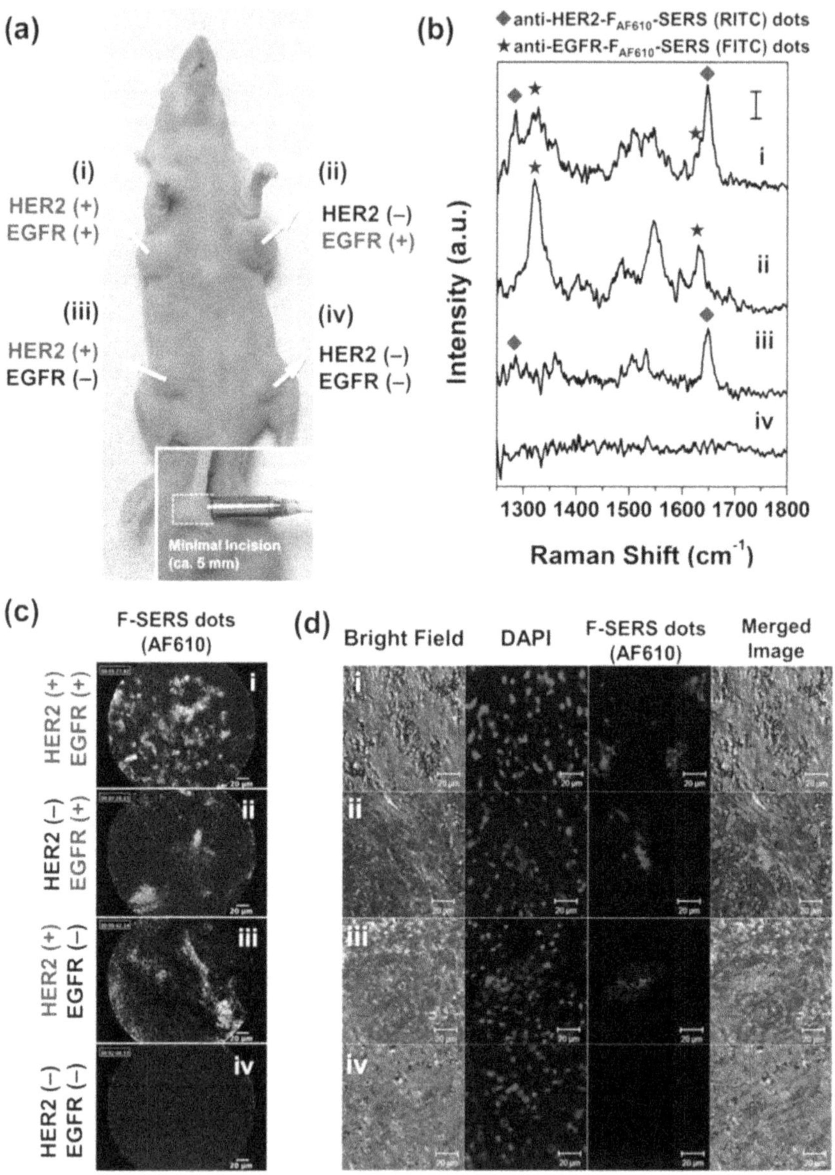

Figure 5. In vivo, multiplexed, endoscopic detection of Fluorescnence-SERS (F-SERS) multimodal BFNPs in breast tumor xenografts. Tumors sites (a) were treated with antibodies for HER2 or EGFR to block specific binding by two flavors of F-SERS BFNPs. SERS spectra (b), fluorescence images obtained in real time using an endoscopic probe (c), and confocal fluorescnce images (d) show that the F-SERS particles specifically bound to tumors expressing both targets without blocking (i), one target when the other is blocked (ii, iii), and no targets when both are blocked (iv). Adapted with permission from ref. (166). Copyright 2015 Nature Publishing Group.

Bringing many of these concepts together, Jeong *et al.* have demonstrated an interesting multi-modal approach, using a 200nm silica particle decorated with Raman-labeled 10nm Ag NPs (*166*). These BFNPs could be made in multiple flavors depending on the Raman reporter used. This construct is then covered in a silica shell, which was conjugated to both a fluorescent dye for multi-modal imaging and antibodies (anti-EGFR and anti-HER2) to target xenograft tumors in a mouse model. Given the fluorescence and SERS response from the probes, a Fluorescence-Raman Endoscopic System (FRES) was used to demonstrate the presence of target particles in the tumors through both spectroscopic modalities via endoscopic measurement (Figure 5). Coupled measurements, with fluorescence for sensitivity and SERS for specificity, may prove to be a way forward in future diagnostics.

Though non-cancerous disease diagnosis has been a target of many researchers working on *in vitro* assays, few examples exist of *in vivo* detection via BFNPs. Intercellular adhesion molecule 1 (ICAM-1) expression by the endothelium is an early and key step in the inflammatory response, making ICAM-1 an important biomarker in a number of diseases (*167*, *168*). Targeting ICAM-1 as a marker for atherosclerosis, McQueenie and coworkers (*169*) showed promising results in mice models prone to the disease. The BFNPs in these experiments were the same as those used previously by Gambhir *et al.* (*144*, *147*), but with the addition of an anti-ICAM-1 antibody. Not only were the researchers able to detect the presence of bound BFNPs in the vasculature when ICAM-1 was over expressed, but also able to directly compare SERS and fluorescence signal-to-noise ratios, with the SERS result comparing favorably.

In Vivo Therapeutics

As with *in vitro* diagnostics, the progression has been made from *in vivo* detection of biomarkers and diagnosis to the prospect of targeted theranostics. As mentioned previously, the field of theranostics has seen great advancements (*130*, *131*), but again, we will choose to focus on studies using BFNPs for targeted imaging and therapeutic treatment delivery, of which reports are limited.

While reports of *in vivo* theranosis exist, very few utilise particles that meet the criteria established herein for BFNPs. One true example of the use of targeted BFNPs has been shown by Conde *et al.* (*170*) Gold particles with a Raman reporter and PEG layer were conjugated to Cetuximab (Erbitux Merck), an antibody-drug conjugate. Cetuximab serves as a cell signal inhibitor by binding to EGFR on the surface of cancer cells (*171*). Similar to other *in vivo* studies, Raman studies confirmed antibody (drug) functionalised particles were preferentially accumulated in xenograpt tumor cells (Figure 6). Of greater interest is the report that the use of the antibody-drug conjugates resulted in dramatic tumor size reduction (~70%). This result brings BFNP theranostics from *in vitro* to *in vivo*, and one step closer to clinical application.

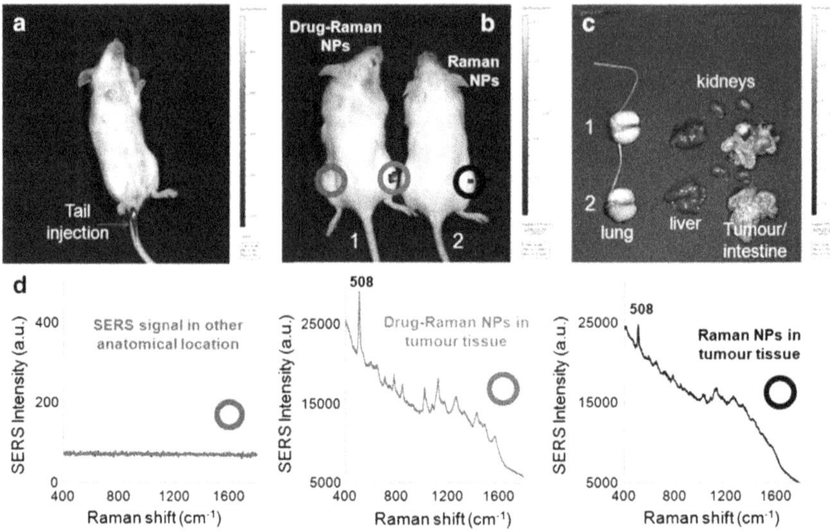

Figure 6. Uptake and distribution of Cetuximab-loaded BFNP in mice. Epi-fluorescence images show emission from DTIC Raman reporter on NP (a) immediately after injection. (b) Drug-loaded BFNPs accumulate in tumor site selectively (1) when compared to non-targeted NP (2). (c) Distribution of BFNPs in mouse lungs, liver and tumors, showing tumor selectivity. (d) SERS spectra collected from tumor sections show much greater SERS signal (~4.5 fold) for drug-loaded NPs (middle) vs. control Raman NPs (right) and a representative control anatomical location (left). Adapted with permission from ref. (170). Copyright 2014 Elsevier B.V.

Conclusions and Forward Outlook

The purpose of this review is to provide an insight in to the recent developments of SERS-active BFNPs and their implementation in clinical analysis. Progress has been made in the areas of nanoparticle synthesis and surface functionalisation utilizing different reporter molecules, encapsulation layers, and biomolecules for clinical diagnostics. These modifications provide SERS-activity, stability, and specificity to BFNPs which are not observed for bare or untargeted NP methods. This positions BFNPs well in the future of diagnostics as nanotechnology in the lab moves toward nanomedicine in the clinic.

It is inevitable that great progress is often met with difficulties; there is still much progress to be made before BFNPs become commonplace in diagnostics and therapeutics. It is clear from the outset that for BFNPs to be used in a clinical environment, neither a top-down (clinicians directing researchers) or bottom-up (researchers directing clinicians) approach to assay design can be taken. Rather, constant communication between researchers and clinicians needs to be implemented to relay important information about the diagnostic tools' needs. Clinicians should inform target choice based on clinical need, while researchers provide a perspective based on their own expertise. Perhaps most vital is that all the people concerned with the project elude to the same vision of the final format.

Another potential issue to address is nanoparticle reproducibility with respect to important, fundamental metrics: size, shape, stability, LSPR, SERS activity and batch-to-batch reproducibility. The consistency demonstrated in particle synthesis does not inherently ensure consistency in SERS performance in a given assay, though encapsulation has been shown to aid in this front. Further, a systematic study of Raman reporters will help to maximize the utility of BFNPs further along in the process.

Rigorous investigation of BFNPs with respect to cell and animal toxicity must also be furthered (*139–142*). Toxicity studies should be careful to be focused on the effects of fully realised BFNPs, with consideration of both the lethal and adverse BFNP doses in the context of what would be a diagnostic or therapeutic dose. There are also concerns to be addressed with delivery routes and nanoparticle clearance once the relevant result is obtained. These issues will need to be addressed for each 'flavor' of BFNPs to achieve regulatory approval for use in humans.

A host of potential issues exist in performing clinically relevant measurements. For each potential application, the appropriate functionalisation strategy and measurement format, such as ELISA, lateral flow, microtitre plate, lab-on-a-chip, spectroscopy or imaging, must be determined through consideration of many factors. Regard should be given to the target analyte concentration and dynamic range, sample matrix (cells, tissue, serum, etc.) and availability, relevant spatial and temporal scales, and measurement sensitivity and specificity. In fact, some clinically relevant targets do not require the sensitivity afforded by SERS assays based on BFNPs. Sensitivity and specificity are important in minimising false clinical results. While SERS of BFNPs provides greater sensitivity and spectral sensitivity for particle differentiation, the specificity of the BFNP-target binding must also be contemplated and rigorously cross-validated to ensure a result's veracity. Sensitivity in these applications isn't only a function of how much signal can be generated from an arbitrary number of successfully targeted BFNPs in the assay, but also how well that signal can be differentiated from a background of spectral and/or non-specific interferents.

Another important consideration is whether the diagnostic tool should be capable of detecting only one analyte or amenable to multiplexing. For example, it could be beneficial to target multiple analytes in a single sample over a broad concentration range rather than targeting a single analyte with the best possible limit of detection. This has the advantage of strengthening the assay from a clinical perspective, where the presence of multiple target analytes may better inform the clinician and the diagnosis.

Further, some measurements may require dedicated Raman instrumentation, specifically those for *in vivo* measurements. If BFNPs are to be of use *in vivo*, it will be important to not only develop particles with tailored LSPRs, but also sensitive instrumentation capable of working well in the NIR-tissue penetration window. Purpose-built instrumentation may be necessary as referenced above (*159–161*), contributing to the cost per assay; however, on the *in* vitro side one could imagine automatic solution- or plate-based systems being used for a multitude of immunoassays. With an eye toward both point of care and the developing world, great strides have been made in developing handheld Raman spectrometers, but further optimisation of both instruments and assays will need to be performed for these methods to reach their potential.

When focussing on patient diagnosis, evaluations of the current market is essential; the new tool, apart from giving faster results, should also show a significant advantage over currently applied methods and comparable cost effectiveness. It must also fit into the everyday diagnostic protocols of a clinic or hospital, where samples are not always collected and immediately evaluated but perhaps stored, analysed and reported within a variable time period.

Finally, the biggest challenge may arise not in synthesis, functionalisation, or instrumentation, but interpretation of the data generated by measurement. As previously mentioned, an assay's sensitivity is determined in part by the presence or absence of background signals. Steps can be taken to mitigate this background through experimental design or statistical tools, but these are generally non-trivial exercises. Advanced statistical tools, such as direct classical and partial least squares and principle component analysis have already advanced the field, and targeted application of these methods and their descendants may serve. These tools will be of significance when multiplexing for the detection of multiple biomarkers is the intended goal. In any event, the data generated must be converted, ideally in real time, into a useful output for the clinician. The goal ought to be to provide information to the clinician so they can work with their patient to create the best outcome. BFNP-based assays will truly take off when they successfully demonstrate value added to clinicians.

References

1. Mitchell, A. L.; Gajjar, K. B.; Theophilou, G.; Martin, F. L.; Martin-Hirsch, P. L. *J. Biophotonics* **2014**, *7*, 153–165.
2. Baker, M. J.; Trevisan, J.; Bassan, P.; Bhargava, R.; Butler, H. J.; Dorling, K. M.; Fielden, P. R.; Fogarty, S. W.; Fullwood, N. J.; Heys, K. A.; Hughes, C.; Lasch, P.; Martin-Hirsch, P. L.; Obinaju, B.; Sockalingum, G. D.; Sule-Suso, J.; Strong, R. J.; Walsh, M. J.; Wood, B. R.; Gardner, P.; Martin, F. L. *Nat. Protoc.* **2014**, *9*, 1771–1791.
3. Bellisola, G.; Sorio, C. *Am. J. Cancer Res.* **2012**, *2*, 1–21.
4. Chen, P.; Shen, A. G.; Zhou, X. D.; Hu, J. M. *Anal. Methods* **2011**, *3*, 1257–1269.
5. Das, R. S.; Agrawal, Y. K. *Vib. Spectrosc.* **2011**, *57*, 163–176.
6. Krafft, C.; Popp, J. *Anal. Bioanal. Chem.* **2015**, *407*, 699–717.
7. Kendall, C.; Isabelle, M.; Bazant-Hegemark, F.; Hutchings, J.; Orr, L.; Babrah, J.; Baker, R.; Stone, N. *Analyst* **2009**, *134*, 1029–1045.
8. Evans, C. L.; Xie, X. S. *Annu. Rev. Anal. Chem.* **2008**, *1*, 883–909.
9. Freudiger, C. W.; Min, W.; Saar, B. G.; Lu, S.; Holtom, G. R.; He, C. W.; Tsai, J. C.; Kang, J. X.; Xie, X. S. *Science* **2008**, *322*, 1857–1861.
10. Saar, B. G.; Freudiger, C. W.; Reichman, J.; Stanley, C. M.; Holtom, G. R.; Xie, X. S. *Science* **2010**, *330*, 1368–1370.

11. Antonio, K. A.; Schultz, Z. D. *Anal. Chem.* **2014**, *86*, 30–46.
12. Schie, I. W.; Krafft, C.; Popp, J. *Analyst* **2015**, *140*, 3897–3909.
13. Lehner, R.; Wang, X.; Marsch, S.; Hunziker, P. *Nanomedicine* **2013**, *9*, 742–757.
14. Stiles, P. L.; Dieringer, J. A.; Shah, N. C.; Van Duyne, R. R. *Annu. Rev. Anal. Chem.* **2008**, *1*, 601–626.
15. Kneipp, K.; Wang, Y.; Kneipp, H.; Perelman, L. T.; Itzkan, I.; Dasari, R.; Feld, M. S. *Phys. Rev. Lett.* **1997**, *78*, 1667–1670.
16. Etchegoin, P. G.; Le Ru, E. C. *Phys Chem Chem Phys* **2008**, *10*, 6079–6089.
17. Kho, K. W.; Fu, C. Y.; Dinish, U. S.; Olivo, M. *J. Biophotonics* **2011**, *4*, 667–684.
18. Ellis, D. I.; Cowcher, D. P.; Ashton, L.; O'Hagan, S.; Goodacre, R. *Analyst* **2013**, *138*, 3871–3884.
19. Yigit, M. V.; Medarova, Z. *Am. J. Nucl. Med. Mol. Imaging* **2012**, *2*, 232–241.
20. Vendrell, M.; Maiti, K. K.; Dhaliwal, K.; Chang, Y. T. *Trends Biotechnol.* **2013**, *31*, 249–257.
21. Laing, S.; Gracie, K.; Faulds, K. *Chem. Soc. Rev.* **2016**, *45*, 1901–18.
22. Zeng, S. W.; Yong, K. T.; Roy, I.; Dinh, X. Q.; Yu, X.; Luan, F. *Plasmonics* **2011**, *6*, 491–506.
23. Bedics, M. A.; Kearns, H.; Cox, J. M.; Mabbott, S.; Ali, F.; Shand, N. C.; Faulds, K.; Benedict, J. B.; Graham, D.; Detty, M. R. *Chem. Sci.* **2015**, *6*, 2302–2306.
24. Moreton, S.; Faulds, K.; Shand, N. C.; Bedics, M. A.; Detty, M. R.; Graham, D. *Nanoscale* **2015**, *7*, 6075–6082.
25. Wang, Y. Q.; Yan, B.; Chen, L. X. *Chem. Rev.* **2013**, *113*, 1391–1428.
26. Wuelfing, W. P.; Gross, S. M.; Miles, D. T.; Murray, R. W. *J. Am. Chem. Soc.* **1998**, *120*, 12696–12697.
27. Mulvaney, S. P.; Musick, M. D.; Keating, C. D.; Natan, M. J. *Langmuir* **2003**, *19*, 4784–4790.
28. Sokolov, K.; Follen, M.; Aaron, J.; Pavlova, I.; Malpica, A.; Lotan, R.; Richards-Kortum, R. *Cancer Res.* **2003**, *63*, 1999–2004.
29. Mirkin, C. A.; Letsinger, R. L.; Mucic, R. C.; Storhoff, J. J. *Nature* **1996**, *382*, 607–609.
30. Tkachenko, A. G.; Xie, H.; Coleman, D.; Glomm, W.; Ryan, J.; Anderson, M. F.; Franzen, S.; Feldheim, D. L. *J. Am. Chem. Soc.* **2003**, *125*, 4700–4701.
31. Medley, C. D.; Smith, J. E.; Tang, Z.; Wu, Y.; Bamrungsap, S.; Tan, W. H. *Anal. Chem.* **2008**, *80*, 1067–1072.
32. Bujacz, A. *Acta Crystallogr., Sect. D: Biol. Crystallogr.* **2012**, *68*, 1278–1289.
33. Schmid, G.; Fenske, D. *Philos. Trans. R. Soc., A* **2010**, *368*, 1207–1210.
34. Campion, A.; Kambhampati, P. *Chem. Soc. Rev.* **1998**, *27*, 241–250.
35. Otto, A.; Mrozek, I.; Grabhorn, H.; Akemann, W. *J. Phys.: Condens. Matter* **1992**, *4*, 1143–1212.
36. Fleischmann, M.; Hendra, P. J.; Mcquilla, Aj *Chem. Phys. Lett.* **1974**, *26*, 163–166.
37. Stockman, M. I. *Top. Appl. Phys.* **2006**, *103*, 47–65.

38. Fan, M. K.; Andrade, G. F. S.; Brolo, A. G. *Anal. Chim. Acta* **2011**, *693*, 7–25.
39. Mabbott, S.; Larmour, I. A.; Vishnyakov, V.; Xu, Y.; Graham, D.; Goodacre, R. *Analyst* **2012**, *137*, 2791–2798.
40. Negri, P.; Dluhy, R. A. *J. Biophotonics* **2013**, *6*, 20–35.
41. Faraday, M. The Bakerian Lecture: Experimental Relations of Gold (and Other Metals) to Light. *Phil. Trans. R. Soc. London* **1857**, *147*, 145–181.
42. Bakrania, S. D.; Rathore, G. K.; Wooldridge, M. S. *J. Therm. Anal. Calorim.* **2009**, *95*, 117–122.
43. Chico, R.; Castillejos, E.; Serp, P.; Coco, S.; Espinet, P. *Inorg. Chem.* **2011**, *50*, 8654–8662.
44. Crespo, J.; Guari, Y.; Ibarra, A.; Larionova, J.; Lasanta, T.; Laurencin, D.; Lopez-de-Luzuriaga, J. M.; Monge, M.; Olmos, M. E.; Richeter, S. *Dalton Trans.* **2014**, *43*, 15713–15718.
45. Kashiwagi, Y.; Yamamoto, M.; Nakamoto, M. *J. Colloid. Interf. Sci.* **2006**, *300*, 169–175.
46. Nakamoto, M.; Kashiwagi, Y.; Yamamoto, M. *Inorg. Chim. Acta* **2005**, *358*, 4229–4236.
47. Santagata, A.; Guarnaccio, A.; Pietrangeli, D.; Szegedi, A.; Valyon, J.; De Stefanis, A.; De Bonis, A.; Teghil, R.; Sansone, M.; Mollica, D.; Parisi, G. P. *J. Phys. D: Appl. Phys.* **2015**, *48*.
48. Sobhan, M. A.; Goldys, E. M. *Nanotechnol. Sci. Technol.* **2010**, 205–223.
49. Sreeja, R.; Reshmi, R.; Aneesh, P. M.; Jayaraj, M. K. *Sci. Adv. Mater.* **2012**, *4*, 439–448.
50. Polte, J.; Erler, R.; Thunemann, A. F.; Sokolov, S.; Ahner, T. T.; Rademann, K.; Emmerling, F.; Kraehnert, R. *ACS Nano* **2010**, *4*, 1076–1082.
51. Polte, J.; Tuaev, X.; Wuithschick, M.; Fischer, A.; Thuenemann, A. F.; Rademann, K.; Kraehnert, R.; Emmerling, F. *ACS Nano* **2012**, *6*, 5791–5802.
52. Li, J.; Wu, J.; Zhang, X.; Liu, Y.; Zhou, D.; Sun, H. Z.; Zhang, H.; Yang, B. *J. Phys. Chem. C* **2011**, *115*, 3630–3637.
53. Vigderman, L.; Zubarev, E. R. *Chem. Mater.* **2013**, *25*, 1450–1457.
54. Ma, Y. Y.; Li, W. Y.; Cho, E. C.; Li, Z. Y.; Yu, T. K.; Zeng, J.; Xie, Z. X.; Xia, Y. N. *ACS Nano* **2010**, *4*, 6725–6734.
55. Qin, Y. Q.; Ji, X. H.; Jing, J.; Liu, H.; Wu, H. L.; Yang, W. S. Size control over spherical silver nanoparticles by ascorbic acid reduction. *Colloids Surf., A* **2010**, *372*, 172–176.
56. Afzal, A. B.; Akhtar, M. J.; Nadeem, M.; Hassan, M. M. *J. Phys. Chem. C* **2009**, *113*, 17560–17565.
57. Bankar, A.; Joshi, B.; Kumar, A. R.; Zinjarde, S. *Colloids Surf., B* **2010**, *80*, 45–50.
58. Nori, N. M.; Abdi, K.; Khoshayand, M. R.; Ahmadi, S. H.; Lamei, N.; Shahverdi, A. R. *J. Exp. Nanosci.* **2013**, *8*, 442–450.
59. Anderson, R.; Buscall, R.; Eldridge, R.; Mulvaney, P.; Scales, P. *RSC Adv.* **2014**, *4*, 46876–46886.
60. Slepicka, P.; Elashnikov, R.; Ulbrich, P.; Staszek, M.; Kolska, Z.; Svorcik, V. *J. Nanopart. Res.* **2015**, *17*, 11.

61. Stevenson, A. P. Z.; Bea, D. B.; Civit, S.; Contera, S. A.; Cerveto, A. I.; Trigueros, S. *Nanoscale Res. Lett.* **2012**, *7*, 1–8.
62. Wu, C.; Guo, H. Y.; Hu, J. W. *Acta Chim Sinica* **2009**, *67*, 1621–1625.
63. Yoo, E. J.; Li, T.; Park, H. G.; Chang, Y. K. *Ultramicroscopy* **2008**, *108*, 1273–1277.
64. Bhumkar, D. R.; Joshi, H. M.; Sastry, M.; Pokharkar, V. B. *Pharm. Res.* **2007**, *24*, 1415–1426.
65. Huang, H. Z.; Yang, X. R. *Carbohyd. Res.* **2004**, *339*, 2627–2631.
66. Leiva, A.; Bonardd, S.; Pino, M.; Saldias, C.; Kortaberria, G.; Radic, D. *Eur. Polym. J.* **2015**, *68*, 419–431.
67. Wei, D. W.; Sun, W. Y.; Qian, W. P.; Ye, Y. Z.; Ma, X. Y. *Carbohyd. Res.* **2009**, *344*, 2375–2382.
68. Andrade, G. F. S.; Temperini, M. L. A. *J. Raman Spectrosc.* **2009**, *40*, 1989–1995.
69. Gu, W.; Fan, X. M.; Yao, J. L.; Ren, B.; Gu, R. A.; Tian, Z. Q. *J. Raman Spectrosc.* **2009**, *40*, 405–410.
70. Li, D. Y.; Ouyang, Y.; Chen, L.; Cao, W. R.; Shi, S. H. *Indian J. Phys.* **2011**, *85*, 293–299.
71. Kim, K.; Kim, K. L.; Lee, H. B.; Shin, K. S. *J. Phys. Chem. C* **2010**, *114*, 18679–18685.
72. Park, Y. K.; Kim, T. H.; Park, S. *J. Mater. Chem.* **2010**, *20*, 3637–3641.
73. Song, H. M.; Deng, L.; Khashab, N. M. *Nanoscale* **2013**, *5*, 4321–4329.
74. Vidal-Iglesias, F. J.; Solla-Gullon, J.; Rodes, A.; Feliu, J. M.; Perez, J. M. *Surf. Sci.* **2015**, *631*, 213–219.
75. Cui, L.; Liu, Z.; Duan, S.; Wu, D. Y.; Ren, B.; Tian, Z. Q.; Zou, S. Z. *J. Phys. Chem. B* **2005**, *109*, 17597–17602.
76. Yang, Z. L.; Wu, D. Y.; Ren, B.; Zhou, H. G.; Tian, Z. Q. *Spectrosc. Spect. Anal.* **2004**, *24*, 682–685.
77. Zhong, Q. L.; Huang, P.; Zhang, B.; Yang, X. Y.; Ding, Y. M.; Zhou, H. H.; Ren, B.; Tian, Z. Q. *Acta Phys.-Chim. Sin.* **2006**, *22*, 291–295.
78. Yin, A. X.; Liu, W. C.; Ke, J.; Zhu, W.; Gu, J.; Zhang, Y. W.; Yan, C. H. *J. Am. Chem. Soc.* **2012**, *134*, 20479–20489.
79. Ren, B.; Lin, X. F.; Yang, Z. L.; Liu, G. K.; Aroca, R. F.; Mao, B. W.; Tian, Z. Q. *J. Am. Chem. Soc.* **2003**, *125*, 9598–9599.
80. Muniz-Miranda, M.; Gellini, C.; Giorgetti, E. *J. Phys. Chem. C* **2011**, *115*, 5021–5027.
81. Shao, Q.; Que, R. H.; Shao, M. W.; Cheng, L.; Lee, S. T. *Adv. Funct. Mater.* **2012**, *22*, 2067–2070.
82. Li, W. H.; Zamani, R.; Gil, P. R.; Pelaz, B.; Ibanez, M.; Cadavid, D.; Shavel, A.; Alvarez-Puebla, R. A.; Parak, W. J.; Arbiol, J.; Cabot, A. *J. Am. Chem. Soc.* **2013**, *135*, 7098–7101.
83. Zhang, D. H.; Yang, H. B. *Phys. B* **2013**, *415*, 44–48.
84. Tiwari, A. D.; Mishra, A. K.; Mishra, S. B.; Kuvarega, A. T.; Mamba, B. B. *Carbohyd. Polym.* **2013**, *92*, 1402–1407.
85. Cobley, C. M.; Chen, J. Y.; Cho, E. C.; Wang, L. V.; Xia, Y. N. *Chem. Soc. Rev.* **2011**, *40*, 44–56.
86. Larguinho, M.; Baptista, P. V. *J. Proteomics* **2012**, *75*, 2811–2823.

87. Smith, A. M.; Mancini, M. C.; Nie, S. M. *Nat. Nanotechnol.* **2009**, *4*, 710–711.
88. Esnouf, A.; Wright, P. A.; Moore, J. C.; Ahmed, S. *Acupuncture Electro.* **2007**, *32*, 81–86.
89. Turkevich, J.; Stevenson, P. C.; Hillier, J. *Discuss. Faraday Soc.* **1951**, *11*, 55.
90. Frens, G. *Kolloid Z. Z. Polym.* **1972**, *250*, 736.
91. Frens, G. *Nature, Phys. Sci.* **1973**, *241*, 20–22.
92. Li, C. F.; Li, D. X.; Wan, G. Q.; Xu, J.; Hou, W. G. *Nanoscale Res. Lett.* **2011**, *6*, 440.
93. Bastus, N. G.; Comenge, J.; Puntes, V. *Langmuir* **2011**, *27*, 11098–11105.
94. Schulz, F.; Homolka, T.; Bastus, N. G.; Puntes, V.; Weller, H.; Vossmeyer, T. *Langmuir* **2014**, *30*, 10779–10784.
95. Brust, M.; Walker, M.; Bethell, D.; Schiffrin, D. J.; Whyman, R. *J. Chem. Soc., Chem. Commun.* **1994**, *7*, 801–802.
96. Lee, P. C.; Meisel, D. *J. Phys. Chem.* **1982**, *86*, 3391–3395.
97. Bastus, N. G.; Merkoci, F.; Piella, J.; Puntes, V. *Chem. Mater.* **2014**, *26*, 2836–2846.
98. Leopold, N.; Lendl, B. *J. Phys. Chem. B* **2003**, *107*, 5723–5727.
99. Orendorff, C. J.; Gearheart, L.; Jana, N. R.; Murphy, C. J. *Phys. Chem. Chem. Phys.* **2006**, *8*, 165–170.
100. Ros, I.; Placido, T.; Amendola, V.; Marinzi, C.; Manfredi, N.; Comparelli, R.; Striccoli, M.; Agostiano, A.; Abbotto, A.; Pedron, D.; Pilot, R.; Bozio, R. *Plasmonics* **2014**, *9*, 581–593.
101. Smitha, S. L.; Gopchandran, K. G.; Ravindran, T. R.; Prasad, V. S. *Nanotechnology* **2011**, *22*, 265705.
102. Jana, N. R.; Gearheart, L.; Murphy, C. J. *Adv. Mater.* **2001**, *13*, 1389–1393.
103. Nikoobakht, B.; El-Sayed, M. A. *Chem. Mater.* **2003**, *15*, 1957–1962.
104. John, C. L.; Strating, S. L.; Shephard, K. A.; Zhao, J. X. *RSC Adv.* **2013**, *3*, 10909–10918.
105. Xu, X. L.; Zhao, Y. Y.; Xue, X. D.; Huo, S. D.; Chen, F.; Zou, G. Z.; Liang, X. J. *J. Mater. Chem. A* **2014**, *2*, 3528–3535.
106. Jana, N. R.; Gearheart, L.; Murphy, C. J. *Chem. Commun.* **2001**, *7*, 617–618.
107. Caswell, K. K.; Bender, C. M.; Murphy, C. J. *Nano Lett.* **2003**, *3*, 667–669.
108. Zhang, Y.; Hong, H.; Myklejord, D. V.; Cai, W. B. *Small* **2011**, *7*, 3261–3269.
109. Gu, X.; Nie, C. G.; Lai, Y. K.; Lin, C. J. *Mater. Chem. Phys.* **2006**, *96*, 217–222.
110. Vigderman, L.; Khanal, B. P.; Zubarev, E. R. *Adv. Mater.* **2012**, *24*, 4811–4841.
111. Vo-Dinh, T.; Fales, A. M.; Griffin, G. D.; Khoury, C. G.; Liu, Y.; Ngo, H.; Norton, S. J.; Register, J. K.; Wang, H. N.; Yuan, H. *Nanoscale* **2013**, *5*, 10127–10140.
112. Chi, X. Q.; Huang, D. T.; Zhao, Z. H.; Zhou, Z. J.; Yin, Z. Y.; Gao, J. H. *Biomaterials* **2012**, *33*, 189–206.
113. Wang, G. F.; Lipert, R. J.; Jain, M.; Kaur, S.; Chakraboty, S.; Torres, M. P.; Batra, S. K.; Brand, R. E.; Porter, M. D. *Anal. Chem.* **2011**, *83*, 2554–2561.

114. Krasnoslobodtsev, A. V.; Torres, M. P.; Kaur, S.; Vlassiouk, I. V.; Lipert, R. J.; Jain, M.; Batra, S. K.; Lyubchenko, Y. L. *J. Nanomed. Nanotechnol.* **2015**, *11*, 167–173.
115. Granger, J. H.; Granger, M. C.; Firpo, M. A.; Mulvihill, S. J.; Porter, M. D. *Analyst* **2013**, *138*, 410–416.
116. Dufek, E. J.; Ehlert, B.; Granger, M. C.; Sandrock, T. M.; Legge, S. L.; Herrmann, M. G.; Meikle, A. W.; Porter, M. D. *Analyst* **2010**, *135*, 2811–2817.
117. Li, M.; Cushing, S. K.; Zhang, J. M.; Suri, S.; Evans, R.; Petros, W. P.; Gibson, L. F.; Ma, D. L.; Liu, Y. X.; Wu, N. Q. *ASC Nano* **2013**, *7*, 4967–4976.
118. Chen, G.; Chen, Y. P.; Zheng, X. W.; He, C.; Lu, J. P.; Feng, S. Y.; Chen, R.; Zeng, H. S. *Appl. Phys. B: Lasers Opt.* **2013**, *113*, 597–602.
119. Chen, Y. P.; Zheng, X. W.; Chen, G.; He, C.; Zhu, W. F.; Feng, S. Y.; Xi, G. Q.; Chen, R.; Lan, F. H.; Zeng, H. S. *Int. J. Nanomed.* **2012**, *7*, 73–82.
120. Wu, L.; Wang, Z. Y.; Zong, S. F.; Chen, H.; Wang, C. L.; Xu, S. H.; Cui, Y. P. *Analyst* **2013**, *138*, 3450–3456.
121. Chon, H.; Lee, S.; Yoon, S. Y.; Chang, S. I.; Lim, D. W.; Choo, J. *Chem. Commun.* **2011**, *47*, 12515–12517.
122. Chon, H.; Lee, S.; Yoon, S. Y.; Lee, E. K.; Chang, S. I.; Choo, J. *Chem. Commun.* **2014**, *50*, 1058–1060.
123. Chon, H.; Wang, R.; Lee, S.; Bang, S. Y.; Lee, H. S.; Bae, S. C.; Hong, S.; Yoon, Y.; Lim, D. W.; deMello, A. J.; Choo, J. *Anal. BioAnal. Chem.* **2015**, *407*, 8353–8362.
124. Balzerova, A.; Fargasova, A.; Markova, Z.; Ranc, V.; Zboril, R. *Anal. Chem.* **2014**, *86*, 11107–11114.
125. Ma, H. T.; Liu, J. P.; Ali, M. M.; Mahmood, M. A. I.; Labanieh, L.; Lu, M. R.; Iqbal, S. M.; Zhang, Q.; Zhao, W. A.; Wan, Y. *Chem. Soc. Rev.* **2015**, *44*, 1240–1256.
126. Sun, H. G.; Zu, Y. *Small* **2015**, *11*, 2352–2364.
127. Toh, S. Y.; Citartan, M.; Gopinath, S. C. B.; Tang, T. H. *Biosens. Bioelectron.* **2015**, *64*, 392–403.
128. He, J. C.; Li, G. K.; Hu, Y. L. *Anal. Chem.* **2015**, *87*, 11039–11047.
129. Davis, M. E.; Chen, Z. G.; Shin, D. M. *Nat. Rev. Drug Discovery* **2008**, *7*, 771–782.
130. Xie, J.; Lee, S.; Chen, X. *Adv. Drug Delivery Rev.* **2010**, *62*, 1064–1079.
131. Lim, E. K.; Kim, T.; Paik, S.; Haam, S.; Huh, Y. M.; Lee, K. *Chem. Rev.* **2015**, *115*, 327–394.
132. Nima, Z. A.; Mahmood, M.; Xu, Y.; Mustafa, T.; Watanabe, F.; Nedosekin, D. A.; Juratli, M. A.; Fahmi, T.; Galanzha, E. I.; Nolan, J. P.; Basnakian, A. G.; Zharov, V. P.; Biris, A. S. *Sci. Rep.* **2014**, *4*, 4752.
133. Wu, X.; Luo, L.; Yang, S.; Ma, X.; Li, Y.; Dong, C.; Tian, Y.; Zhang, L.; Shen, Z.; Wu, A. *ACS Appl. Mater. Interfaces* **2015**, *7*, 9965–9971.
134. Song, J.; Pu, L.; Zhou, J.; Duan, B.; Duan, H. *ACS Nano* **2013**, *7*, 9947–9960.
135. Boca-Farcau, S.; Potara, M.; Simon, T.; Juhem, A.; Baldeck, P.; Astilean, S. *Mol. Pharm.* **2014**, *11*, 391–399.

136. Chen, Z.; Yu, D.; Huang, Y.; Zhang, Z.; Liu, T.; Zhan, J. *Sci. Rep.* **2014**, *4*, 6709.
137. Wu, P.; Gao, Y.; Lu, Y.; Zhang, H.; Cai, C. *Analyst* **2013**, *138*, 6501–6510.
138. Fales, A. M.; Yuan, H.; Vo-Dinh, T. *Mol. Pharm.* **2013**, *10*, 2291–2298.
139. Murphy, C. J.; Gole, A. M.; Stone, J. W.; Sisco, P. N. *Acc. Chem. Res.* **2008**, *41*, 1721–1730.
140. Alkilany, A. M.; Murphy, C. J. *J. Nanopart. Res.* **2010**, *12*, 2313–2333.
141. Zavaleta, C. L.; Hartman, K. B.; Miao, Z.; James, M. L.; Kempen, P.; Thakor, A. S.; Nielsen, C. H.; Sinclair, R.; Cheng, Z.; Gambhir, S. S. *Small* **2011**, *7*, 2232–2240.
142. Lasagna-Reeves, C.; Gonzalez-Romero, D.; Barria, M. A.; Olmedo, I.; Clos, A.; Sadagopa Ramanujam, V. M.; Urayama, A.; Vergara, L.; Kogan, M. J.; Soto, C. *Biochem. Biophys. Res. Commun.* **2010**, *393*, 649–655.
143. Qian, X.; Peng, X. H.; Ansari, D. O.; Yin-Goen, Q.; Chen, G. Z.; Shin, D. M.; Yang, L.; Young, A. N.; Wang, M. D.; Nie, S. *Nat. Biotechnol.* **2008**, *26*, 83–90.
144. Jokerst, J. V.; Miao, Z.; Zavaleta, C.; Cheng, Z.; Gambhir, S. S. *Small* **2011**, *7*, 625–633.
145. Samanta, A.; Maiti, K. K.; Soh, K. S.; Liao, X.; Vendrell, M.; Dinish, U. S.; Yun, S. W.; Bhuvaneswari, R.; Kim, H.; Rautela, S.; Chung, J.; Olivo, M.; Chang, Y. T. *Angew. Chem., Int. Ed. Engl.* **2011**, *50*, 6089–6092.
146. Harmsen, S.; Bedics, M. A.; Wall, M. A.; Huang, R.; Detty, M. R.; Kircher, M. F. *Nat. Commun.* **2015**, *6*, 6570.
147. Zavaleta, C. L.; Smith, B. R.; Walton, I.; Doering, W.; Davis, G.; Shojaei, B.; Natan, M. J.; Gambhir, S. S. *Proc. Natl. Acad. Sci. U.S.A.* **2009**, *106*, 13511–13516.
148. Maiti, K. K.; Dinish, U. S.; Fu, C. Y.; Lee, J. J.; Soh, K. S.; Yun, S. W.; Bhuvaneswari, R.; Olivo, M.; Chang, Y. T. *Biosens. Bioelectron.* **2010**, *26*, 398–403.
149. Maiti, K. K.; Dinish, U. S.; Samanta, A.; Vendrell, M.; Soh, K. S.; Park, S. J.; Olivo, M.; Chang, Y. T. *Nano Today* **2012**, *7*, 85–93.
150. Dinish, U. S.; Balasundaram, G.; Chang, Y. T.; Olivo, M. *Sci. Rep.* **2014**, *4*, 4075.
151. Sinha, L.; Wang, Y.; Yang, C.; Khan, A.; Brankov, J. G.; Liu, J. T.; Tichauer, K. M. *Sci. Rep.* **2015**, *5*, 8582.
152. Wang, Y. W.; Khan, A.; Som, M.; Wang, D.; Chen, Y.; Leigh, S. Y.; Meza, D.; McVeigh, P. Z.; Wilson, B. C.; Liu, J. T. *Technology (Singap World Sci)* **2014**, *2*, 118–132.
153. Matousek, P. *Chem. Soc. Rev.* **2007**, *36*, 1292–1304.
154. Stone, N.; Kerssens, M.; Lloyd, G. R.; Faulds, K.; Graham, D.; Matousek, P. *Chem. Sci.* **2011**, *2*, 776–780.
155. Sharma, B.; Ma, K.; Glucksberg, M. R.; Van Duyne, R. P. *J. Am. Chem. Soc.* **2013**, *135*, 17290–17293.
156. Zavaleta, C. L.; Garai, E.; Liu, J. T.; Sensarn, S.; Mandella, M. J.; Van de Sompel, D.; Friedland, S.; Van Dam, J.; Contag, C. H.; Gambhir, S. S. *Proc. Natl. Acad. Sci. U.S.A.* **2013**, *110*, E2288–E2297.

157. Garai, E.; Sensarn, S.; Zavaleta, C. L.; Loewke, N. O.; Rogalla, S.; Mandella, M. J.; Felt, S. A.; Friedland, S.; Liu, J. T.; Gambhir, S. S.; Contag, C. H. *PLoS One* **2015**, *10*, e0123185.
158. Wang, Y. W.; Kang, S.; Khan, A.; Bao, P. Q.; Liu, J. T. *Biomed. Opt. Express* **2015**, *6*, 3714–3723.
159. Bohndiek, S. E.; Wagadarikar, A.; Zavaleta, C. L.; Van de Sompel, D.; Garai, E.; Jokerst, J. V.; Yazdanfar, S.; Gambhir, S. S. *Proc. Natl. Acad. Sci. U.S.A.* **2013**, *110*, 12408–12413.
160. McVeigh, P. Z.; Mallia, R. J.; Veilleux, I.; Wilson, B. C. *J. Biomed. Opt.* **2013**, *18*, 046011.
161. Karabeber, H.; Huang, R. M.; Iacono, P.; Samii, J. M.; Pitter, K.; Holland, E. C.; Kircher, M. F. *ACS Nano* **2014**, *8*, 9755–9766.
162. Mohs, A. M.; Mancini, M. C.; Singhal, S.; Provenzale, J. M.; Leyland-Jones, B.; Wang, M. D.; Nie, S. *Anal. Chem.* **2010**, *82*, 9058–9065.
163. Bardhan, R.; Chen, W.; Bartels, M.; Perez-Torres, C.; Botero, M. F.; McAninch, R. W.; Contreras, A.; Schiff, R.; Pautler, R. G.; Halas, N. J.; Joshi, A. *Nano Lett.* **2010**, *10*, 4920–4928.
164. Kircher, M. F.; de la Zerda, A.; Jokerst, J. V.; Zavaleta, C. L.; Kempen, P. J.; Mittra, E.; Pitter, K.; Huang, R.; Campos, C.; Habte, F.; Sinclair, R.; Brennan, C. W.; Mellinghoff, I. K.; Holland, E. C.; Gambhir, S. S. *Nat. Med.* **2012**, *18*, 829–834.
165. Dinish, U. S.; Song, Z. G.; Ho, C. J. H.; Balasundaram, G.; Attia, A. B. E.; Lu, X. M.; Tang, B. Z.; Liu, B.; Olivo, M. *Adv. Funct. Mater.* **2015**, *25*, 2316–2325.
166. Jeong, S.; Kim, Y. I.; Kang, H.; Kim, G.; Cha, M. G.; Chang, H.; Jung, K. O.; Kim, Y. H.; Jun, B. H.; Hwang, D. W.; Lee, Y. S.; Youn, H.; Lee, Y. S.; Kang, K. W.; Lee, D. S.; Jeong, D. H. *Sci. Rep.* **2015**, *5*, 9455.
167. Anderson, M. E.; Siahaan, T. J. *Peptides* **2003**, *24*, 487–501.
168. Johnson, P. A.; Alexander, H. D.; McMillan, S. A.; Maxwell, A. P. *Clin. Exp. Immunol.* **1997**, *108*, 234–242.
169. McQueenie, R.; Stevenson, R.; Benson, R.; MacRitchie, N.; McInnes, I.; Maffia, P.; Faulds, K.; Graham, D.; Brewer, J.; Garside, P. *Anal. Chem.* **2012**, *84*, 5968–5975.
170. Conde, J.; Bao, C.; Cui, D.; Baptista, P. V.; Tian, F. *J. Controlled Release* **2014**, *183*, 87–93.
171. Bokemeyer, C.; Van Cutsem, E.; Rougier, P.; Ciardiello, F.; Heeger, S.; Schlichting, M.; Celik, I.; Kohne, C. H. *Eur. J.Cancer* **2012**, *48*, 1466–1475.

Chapter 8

Surface-Enhanced Raman Spectroscopy for the Characterization of Semiconductor Nanostructure Surfaces

Xiaowei Li, Hiro Minamimoto, Satoshi Yasuda, and Kei Murakoshi*

Department of Chemistry, Faculty of Science, Hokkaido University, Sapporo, 060-0810, Japan
***E-mail: kei@sci.hokudai.ac.jp**

During the past few decades, SERS active substrates have been developed significantly. Besides coinage metals and their alloys that have been extensively employed owing to their large enhancement through localized surface plasmon resonance (LSPR), semiconductor QDs have also been validated the enhancement with a factor on the order as high as 10^3 to 10^6 times as SERS active substrates and induced increasing attention due to the widespread application in both SERS spectroscopy and material fields. Semiconductor nanostructures offer a considerable advantage of a much variety of tunable properties, such as band gap, exciton Bohr radius and geometry though various synthetic techniques. In this review, recent SERS studies including enhancement on the lattice vibrations, molecule detection and photochemical reactions by coupling semiconductor nanostructures with LSPR are introduced. We expect such frontier introduction could provide directions to further investigation of semiconductor-metal coupling systems.

To date, surface plasmon resonance has attracted numerous research interest because the oscillation of free electrons in noble metals by incident light generates strong confined electromagnetic field in the vicinity of metal with an enhancement factor reaching up to a scale of 1,000. The discovery of surface-enhanced Raman spectroscopy/scattering (SERS) has opened a new avenue for molecule detection thanks to the breakthrough in limited sensitivity of Raman spectroscopy. Plasmonic surfaces enhance the Raman scattered light from molecules on or near the surface by 10^8 to 10^{12} times (*1*). SERS spectroscopy has been widely employed for chemistry (*2*), biology (*3*) and medicine (*4*, *5*), by utilizing various nanoparticle size, shape, and spacing parameters, and massively parallel.

Based on quantum confinement, semiconductor nanostructures, besides noble metal, also have a considerable advantage of a much variety of tunable properties such as size-dependent bandgap energy, broaden absorption range, and diverse geometries though simple synthetic techniques. Semiconductor nanostructures, especially quantum dots (QDs) have been validated Raman scattering with an enhancement factor on the order as high as 10^3-10^6 times (*6–8*) as SERS active substrates for both lattice vibrations of crystal nanostructures as well as molecule detection, thus SERS studies of semiconductor nanostructures induce increasing attention owing to the widespread application in both vibrational spectroscopy and material fields.

A hybrid system that a plasmonic nanostructure coupled with quantum emitters, such as fluorescent molecules or semiconductor quantum dots (QDs), is an important issue in nano-spectroscopy and nano-optics, involving a wide range of applications, including surface- (*9–11*) or tip-enhanced optical spectroscopy (*12*, *13*), enhanced efficiencies of solar energy harvesting (*14–17*), light-emitting (*18–20*) and sensing (*21–23*). Recent studies have revealed that electromagnetic field of localized surface plasmon resonance (LSPR) is also able to modify absorption of semiconductors and even break optical transitions. Alivisatos's group has proved the breakdown of absorption transitions of CdSe nanorods in the vicinity of LSPR (Figure 1) (*24*); our group previously demonstrated the validity of highly confined electromagnetic field in the gap of Au nanodimer to modify the electronic excitation process of an isolated single-walled carbon nanotube owing to breaking the symmetric selection rule (*25*) (Figure 1).

Vibration involved processes, such as intraband relaxation via phonon scattering (*26*, *27*) as well as exciton-vibrational coupling between excited colloidal QDs and capping ligands (*28*, *29*) do influence exciton recombination processes. In this sense, to investigate semiconductor QDs in strong electromagnetic field of LSPR by SERS will reveal hidden information that cannot be observed by traditional Raman scattering, and thus provide insight to better understanding intrinsic photophysical properties and benefit the design and development of optical devices such as photo energy switch and light emitting diode.

Benefiting from semimetal properties such as large dielectric constant and relatively small effective mass of electrons and holes, lead sulfide (PbS) is easily to present strong quantum confinement and becomes an excellent candidate materials for device applications, especially for photo conversion devices based on multiple exciton generation (MEG) (*30*, *31*). Not only as an efficient photoenergy conversion material, PbS QDs also perform a noticeable role in SERS detection both by physical (*32, 33*) and chemical (*34*) mechanism. In this review, SERS on semiconductor QDs, especially PbS QDs, will be introduced on the aspect involving a brief introduction of phonon confinement in size-controlled semiconductor nanostructures and phonon bottleneck breaking, a summary of phonon modes detected by SERS and experimental results of molecule characterization by QDs, and determination of electrochemical reactions by QDs. We expect such frontier introduction could provide a view to widely grasp the promising effect between semiconductor QDs and LSPR.

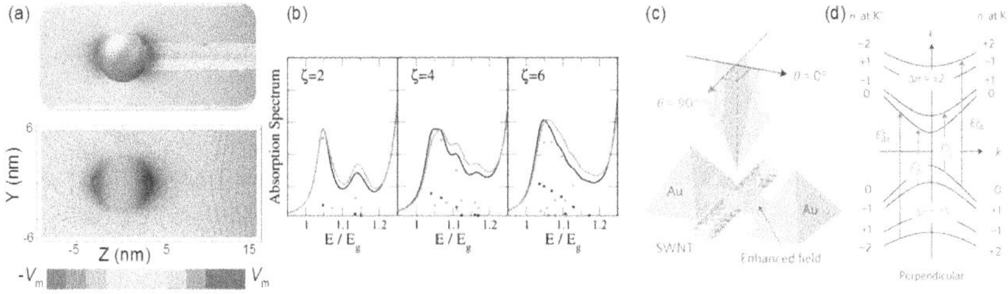

Figure 1. (a) Schematic illustration of the geometry of an Au-tipped CdSe nanorod (upper) and isopotential contour lines of the near-field potential generated by a gold nanoparticle of 6 nm diameter (lower). (b) The absorption rate of a series of CdSe nanorods of different aspect ratios (ξ=2, 4 and 6) for a diameter of D≈2.8 nm normalized to the first absorption peak height without a metal tip and with a 5 nm metal tip. Reproduced with permission from reference (24). Copyright 2012 National Academy of Sciences. (c) Illustration of an SWNT lying in the nanogap of a dimer and the enhanced field polarization. (d) Energy bands and optical transitions of a semiconducting SWNT with allowed transitions of $\Delta n=\pm1$ ($E^S{}_{12}$ and $E^S{}_{21}$) and forbidden transitions of $\Delta n=\pm2$ ($E^S{}_{14}$ and $E^S{}_{41}$). Reproduced with permission from reference (25). Copyright 2013 Nature Publishing Group.

Phonon Confinement and Breaking Phonon Bottleneck in Semiconductor QDs

Absorption of photon with energy exceeds semiconductor's bandgap excite semiconductor to generate bound electron-hole pair known as an exciton. When the initial photon energy is large enough to overcome the exciton binding energy, exciton dissociate into an electron-hole pair each with excess energy above the respective energy bandedge and release the excess energy, the intraband relaxation, as phonon emission. A phonon is a quantum mechanical description of an elementary vibrational motion in which a lattice of atoms or molecules uniformly oscillates at a single frequency (*35*). The relationship that phonon energy (or frequencies, ω) as a function of their wavevector q is called phonon dispersion ω (q) in the Brillouin zone.

Phonon modes of diverse bulk semiconductor have been widely studied by Raman spectroscopy several decades ago. Recent research interest has been focused on semiconductor QDs detection by resonance Raman or SERS, as the size of quantized nanostructure deeply influence on optical and electrical properties. In a quantized regime, not only electron and hole are confined, but also phonon exhibit distinct quantized characteristics which can be detected by Raman spectroscopy or SERS.

The periodicity of the crystal is interrupted when the grain size of semiconductor is smaller than typically 20 lattice parameters because of quantum confinement. The phonon in an isolated grain can get reflected from the boundaries and remain confined within the grain, named phonon confinement. Optical as well as acoustical phonons get confined within the particle. The consequence of phonon confinement is noticeable in the vibrational spectra. In bulk semiconductor, based on the wavevector selection rule of visible Raman scattering, Raman spectral of can only originate from the phonons at $q \approx 0$ and the Raman spectral peaks are the narrow spectral lines at bulk frequency of ω_1 or ω_2 (Figure 2). But in nanostructured crystal, the wavevector uncertainty leads to the relaxation of the selection rule of $q \approx 0$ in visible Raman scattering, so that the phonons in the region of wavevector Δq can join the Raman scattering process, resulting in broadened spectral bands linewidth (*36*). In the case that the dispersion relation of the phonons in the Brillouin zone is negative (Figure 2a, ω_2), phonons of smaller energy dominate in the scattering process, leading to red-shifted Raman shift (Figure 2b, ω_2). Otherwise in a positive dispersion relation, relaxation of momentum Δq results in broaden towards high energy as blue-shifted Raman shift (Figure 2, ω_1).

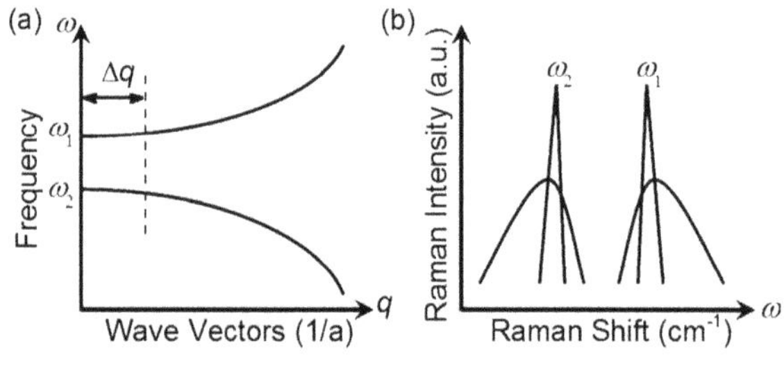

Figure 2. (a) Typical dispersion curves of bulk semiconductors and (b) expected Raman spectral features of nanostructured material.

Previous reported typical Raman spectra of phonon modes in semiconductor nanocrystals, such as GaAs (*37*), CdSe (*38*) and PbS (*39*), are for relatively large sizes which preserve crystalline structure of the bulk crystal. However, as the size reduce to some extent, electronic states in strong confined nanostructures, especially 3-dimentional confined QDs, become sparse with gaps of 100 meV or more, which is multiple of typical optical phonon energy (e.g. ~30 meV for CdSe (*40*) or ~17 meV for PbSe (*41*)), so that intraband relaxation assisted by phonon emission have to involve simultaneous multiple phonon or the participation of very high-energy phonon modes such as surface optical modes (SO). This so-called phonon bottleneck (*42*) make the characterization of size-dependent phonon in QDs with a sparse loading amount become difficult by normal Raman spectroscopy unless under photooxidation (*43*) or resonant condition based on exciton-phonon coupling (*44–47*).

The dipolar electric field of the exciton or electron couples to optical vibrations through the Fröhlich interaction is the so-called exciton- or electron-phonon coupling (*48*, *49*), and the overtone, for example, 2LO of the fundamental 1LO, is the measure of electron-phonon interaction in semiconductors (*50*). To overcome the phonon bottleneck, one solution is efficient multiphonon emission triggered by electron-phonon interactions under resonant condition which is to suitably match the energy between bandgap and excitation photon. Klimov's group has reported that by suppressing efficient electron-hole energy transfer, PbSe QDs exhibit strong size-dependent electron-phonon coupling and intraband relaxation (*51*); Minnaert's group has revealed that multiple phonon-assisted photoluminescence of InAs/GaAs QDs (*52*). Thermal fluctuations of the surface atoms also improve multiphonon resonant Raman scattering because of electron-phonon coupling. By well controlling the resonance condition, laser-heating induced multiphonon emission has been observed for Si nanotubes (*53*), CdS nanocrystals (*54*), and the influence of thermal annealing process on multiphonon scattering of $Cd_{1-x}Mn_xS$ QDs has been discussed (*55*). A. G. Milekhin and co-workers (*56*) obtained LO modes up to the third order for ultra-small CdS QDs under resonant excitation by well controlling the temperature. A. V. Baranov and co-workers (*57*) observed thickness dependency of LO modes in CdSe nanoplates by resonant Raman instead of off-resonant Raman, and ascribed to interaction between electron and phonon propagated in the confined perpendicular direction. These evidences demonstrate the validity of breaking phonon bottleneck by electron-phonon coupling. The vibrations and electronic excitations (electron-phonon coupling) in semiconductor nanostructures influences many photophysical processes and plays a crucial role in the design and development of photoelectric or photo emitting devices, however, current discrepancy between experimental and theoretical studies of size-dependent electron-phonon coupling in semiconductor nanostructures makes this critically important issues a point of debate (*58*, *59*).

Other factors to break the perfect symmetry of the QDs could also overcome the phonon bottleneck. One is surface reconstruction by sufficient coupling between the excited electrons and surface molecules (*60*) according to a recent theoretical calculation (*61*) and experimental results of unexpected fast decay rate of intraband relaxation in PbSe QDs (*41*) with by time-resolved transient absorption measurements. Another fast relaxation pathway competing with phonon cooling is MEG, a process that one high-energy photon generates more than two electron-hole pairs which is promising for photoenergy conversion (*31*, *62*). Confinement enhanced Auger mechanisms has been evidenced to circumvent phonon bottleneck in CdSe QDs (*63*). In addition, highly excited electron-hole pairs in QDs can also result in the absence of phonon bottleneck. Jonas's group has proved that highly excited PbS QDs by phonon energy exceeding the threshold of MEG (~3 E_g) are essentially bulk-like so that the independent hot electrons and holes confine scattering length less than QD's size and benefit rapid phonon scattering (*64*); and Zhu's research confirmed the optical transition breakdown by a bulk-like hot electron-hole pair based on transient Stark effect (*65*).

Although the expected inhibition of intraband relaxation in semiconductor QDs have been proved through various techniques including pump-probe measurements and resonant Raman scattering, experimental investigations by normal Raman spectroscopy are extremely limited. As localized surface plasmon resonance provides strong electromagnetic field which benefit photon trapping in semiconductors, coupling semiconductor nanostructures to LSPR has been proposed to reveal the phonon modes of semiconductor nanostructures.

Phonon Modes Detected by SERS and Mechanisms of Enhancement

Not only vibrational modes of diverse molecules have been investigated by using plasmonic structures, recently, but phonon modes in various semiconductor QDs have been explored and enhanced by SERS. The enhanced mechanism of SERS is normally attributed to electromagnetic through interaction with surface plasmon excitations and chemical (CM) enhancement through charge transfer interaction of the absorbed molecule with the metal surface (*9*, *66*, *67*), in which the former term is responsible for the major portion of SERS enhancement.

Electromagnetic Mechanism (EM)

Electromagnetic enhancement is owing primarily to LSPR. Metals can be considered as a plasma of electrons composed of polarizable, free electrons (mobile), and a positive ion core (immobile). Interaction of the electromagnetic wave with a metal causes the free electrons to coherently oscillate at the plasmon frequency (ω_p) against the immobile positive ion lattice. Although the increase in the local electric field is usually modest, the enhancement in the inelastically scattered light intensity scales to the fourth power, causing a remarkable SERS effect. However, the enhancement effects are highly localized and decay rapidly as the separation between the analyte and the metal particles increases, making SERS a truly surface-sensitive technique.

Besides nano-structured metals that used to generate 'hot-spots', i.e. the localized confinement of high electromagnetic fields in small spaces for SERS, there has been a growing interest in theoretical (*68*) and experimental studies (*69*) of systems that metal nanoparticle (NP) coupled with dielectric semiconductors. High field enhancements for SERS have been observed from various non-metallic materials, such as metal NPs and PbS QDs (*34*). Although PbS QDs do not possess a plasmon resonance by themselves, the high refractive index (*70*) make their presence increase the electric field in the gap between the QDs and metal NPs significantly, because the fact that the charge oscillations of the metallic NP are screened by those induced in the dielectric media (*71*), resulting in localized hybridized NP plasmon confined to the gap between them. Finite-difference time-domain (FDTD) calculations (*32*) have investigated that coupling PbS QDs with noble metal NPs will enhance the electric-field generated by LSPR, and the enhancement is strongly dependent on their sizes and distance, as shown in Figure 3. A recent study by Murakoshi's group (*33*) has applied such electromagnetic enhancement and revealed that coupling PbS QDs with Au NPs is able to improve photoelectric conversion efficiency in an expanded visible light region by well tuning QDs' size and the distance from Au NPs.

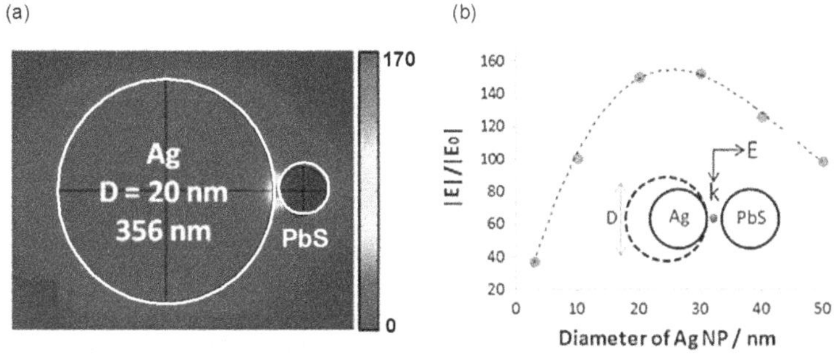

Figure 3. Electric field in the gap between a PbS QD (D= 4.4 nm) placed 0.6 nm away from an Ag NP (D= 20 nm) with the light polarization parallel to the dimer axis excited by 356 nm (left) and the electric field in the gap dependent on the Ag NP size (right). Reproduced with permission from reference (32). Copyright 2013 Wiley. Slightly modified for clearness.

The electromagnetic field generated by LSPR of metal nanostructures such as Ag, Au or Cu provide wide application as light harvesting antenna not only for the design of photoenergy conversion or light emitting devices, but also for modification in detection of molecule or semiconductor nanoparticles by vibrational spectroscopy. Enhanced phonon modes of diverse semiconductors nanostructures have been investigated for CuS (*72*), CdS (*73*), SiO_2 gel (*74*) on plasmonic nanostructures. Phonon modes of ZnO nanocrystals (absorption onset at 350 nm) were observed by Gaponenko and co-workers (*75*) by using Ag nanoparticles under nonresonant excitation of 633 nm (Figure 4) with enhancement up to 10,000 times of the LO phonon mode at 570 cm^{-1} owing to the electromagnetic mechanism by LSPR.

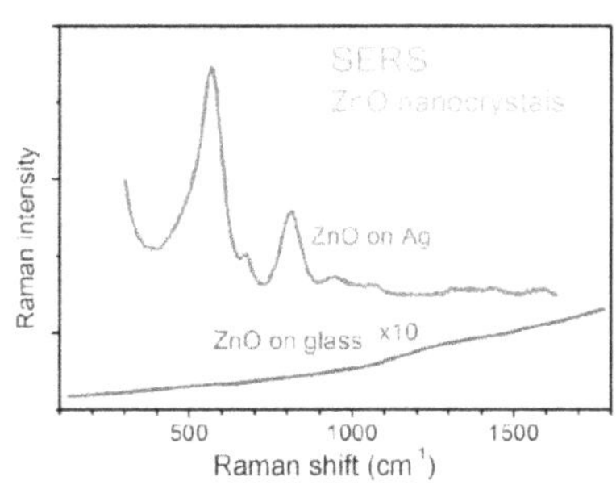

Figure 4. SERS of ZnO nanocrystals on Ag nanoparticles coated glass substrate (ZnO on Ag), and ZnO nanocrystals on glass. Reproduced with permission from reference (75). Copyright 2013 American Chemical Society.

Excitation condition is reasonably essential for spectrum characterization of a semiconductor-noble metal system. The effects of LSPR on optical properties of QD semiconductor were revealed by J.J. Baumberg by using tunable excitation wavelengths (*76*). CdSe/ZnS core/shell QDs (emission at 640 nm) were attached to gold-coated inverted pyramids substrate (Klarite surface, resonance at both 785 and 633 nm). SERS under excitation of 785 nm resulted in molecule vibrations and LO phonon mode of CdSe QDs because of the plasmonic Au nanostructure substrate. Resonant SERS (SERRS) under excitation of 633 nm achieved further enhancement of LO mode and the observation of its overtone 2LO because of the laser is in resonance with both the QDs as well as the plasmonic substrate. A 2-order enhancement was yielded compared to SERS at 785 nm, owing to the combined electronic and plasmonic enhancement. Resonance Raman scattering (RRS) at 514 nm brought about strongly observed overtone 2LO because of electron-phonon coupling.

D. Sigle, J.J. Baumberg and co-workers (*77*) investigated forbidden LO phonon modes of CdSe nanoplates by using highly confined gap mode of plasmon resonance. Ultra-thin individual CdSe nanoplatelets was well fabricated in the gap between a gold nanoparticle with a diameter of 100 nm and a 70-nm-thick gold film. Strongly localized electromagnetic field was confined in the gap, revealing highly confined phonon modes of out-of-plane mode (Figure 5) that is doublet arising from oscillations perpendicular to the platelet and cannot be observed with normal Raman spectroscopy. Varying the thickness of CdSe nanoplates modified the highly localized electromagnetic field and enabled to analyze the Raman shift of phonon dispersion deep into the Brillouin zone. Such work demonstrated the validity of observing new optical properties and breaking selection rules by coupling semiconductors with LSPR.

SERS is a nondestructive tool to investigate the structural evolution of interfaces in core/shell semiconductor QDs. A recent SERS research reported a heterostructures CdSe-based core/shell QDs investigated on Ag nanostructures excited by 514 nm (*78*), giving interface formation between different materials in the QDs. In the case of CdSe/CdS structure, CdSe core exhibits a longitudinal optical mode (LO1) centered at ~209 cm^{-1} (A1 band) \and a surface mode at ~200 cm^{-1} (A2 band); the CdS shell also presents a longitudinal optical mode (LO1) centered at ~290 cm^{-1} (B1 band) and a surface mode at ~275 cm^{-1} (B2 band), as illustrated in Figure 6a~c. A1 and B1 peak frequencies shows a gradual blue-shift with increased shell growth because of the variation of the bond length in the CdSe and CdS lattices. A2 bands first strongly increases and then get saturated as the CdS shell thickness exceeds 3 MLs, and B2 gradually increase with shell growth, arising from not only the surface optical modes of CdSe or CdS, but also the formation of a $CdSe_xS_{1-x}$ alloy between CdSe and CdS (the orange layers in Figure 6g). In the CdSe/$Cd_{0.5}Zn_{0.5}S$ case (Figure 6d~f), the CdS-"like" signal at ~290 cm^{-1} is due to the presence of CdS in the shell composition. B1 bands exhibit a slight increase while the A1, A2, and B2 bands remain almost constant during shell growth, indicating an abrupt interface (the purple layers in Figure 6g) between the CdSe core and the CdZnS shell is confined to the first shell ML (~0.3 nm) . The abrupt interface induces significant stress across the nanocrystals that modifies its bond length.

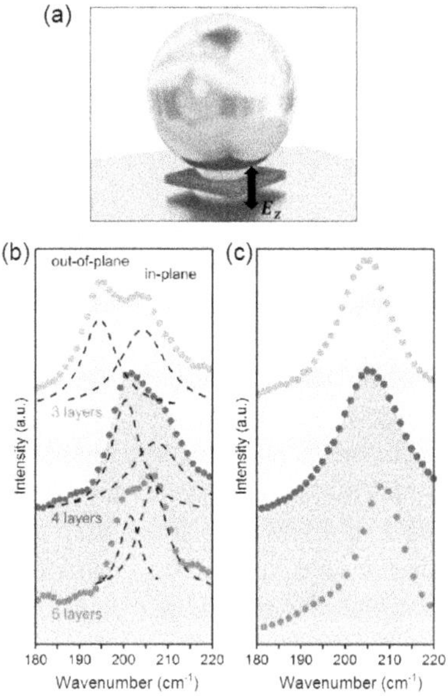

Figure 5. (a) Individual CdSe platelet in a nanogap (b) SERS of CdSe nanoplatelets with different thickness: three, four, and five atomic layers. (c) Conventional Raman scattering of a bulk film of the same platelets. Reproduced with permission from reference (77). Copyright 2014 American Physical Society.

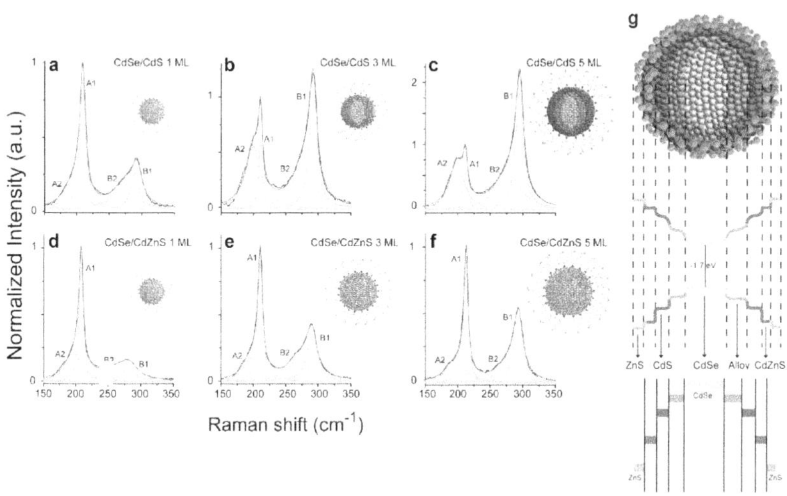

Figure 6. SERRS spectra of CdSe and CdS-like bands for different numbers of ML of CdSe/CdS (a~c) and CdSe/CdZnS QDs (d~f). (g) Schematic structure of CdSe/CdS/CdZnS/ZnS graded QDs with the respective material bulk energy levels. The intermixed layer in the CdSe/CdS/CdZnS/ZnS structure represents the CdSexS1-x alloy formation. Reproduced with permission from reference (78). Copyright 2013 American Chemical Society.

These studies confirmed SERS as an effective approach in understanding the intrinsic properties of semiconductor QDs, which could provide insight to develop the photocatalytic, photoelectric and photoluminescent-based applications.

Chemical Mechanism (CM)

Chemical effect is another independent but relatively weak mechanism that enhances the Raman-scattering cross section of molecule adsorbed on the metal or semiconductor surface, reaching two orders in the enhancement. The proximity between the metal or semiconductor and the analyte molecules causes electronic coupling by charge transfer (CT), resulting in resonant intermediate states as shifted or broadened Raman bands and thereby enhancing Raman scattering (*68*, *69*). Derivation from a metal-molecule system which CT occurs between a Fermi level of metal and HOMO or LUMO of molecule, a semiconductor-molecule system involves conduction or valence band edge, so CT always coupled to a band gap exciton resonance or molecular resonance (*79*, *80*), as shown in Figure 7.

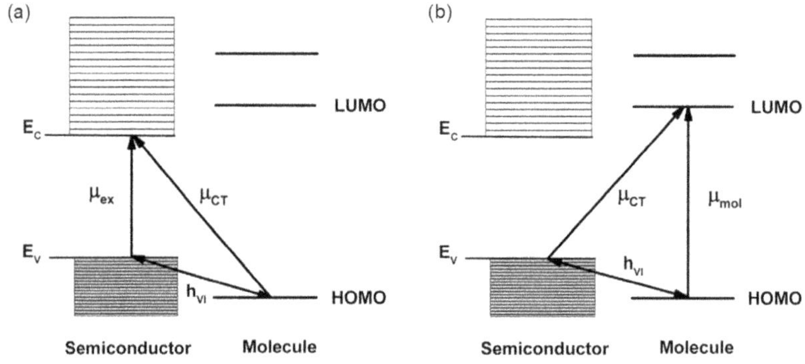

Figure 7. The charge-transfer process between semiconductor and molecule: CT resonance is coupled to (a) exciton resonance and (b) the molecular resonance. Reproduced with permission from reference (80). Copyright 2014 American Chemical Society.

Early on, chemical enhancements of molecule have been widely reported on various bulk semiconductor substrates. For semiconductor nanostructures, the first was by Lucia G. Quagliano (*81*) in which a new vibrational band of pyridine molecule chemisorbed on InAs/GaAs QDs was observed and ascribed to charge transfer from lone pair electrons of the N atom in pyridine (as the donor) to the semiconductor surface (as the acceptor). Since then, charge transfer has been confirmed in molecules adsorbed on diverse semiconductor QDs such as pyridine on colloidal CdSe/CdBeZnSe QDs (*82*), dopamine (*83*) or 4-MBA on TiO_2 (*84*), 4-mercaptopyridine (4-MPy) on ZnSe (*85*), and Nile red on CuTe nanocubes (*86*). Based on quantum confinement, SERS spectra of 4-MPy as capping ligand surrounding PbS QDs was investigated by Lombardi and co-workers (*34*) as a function of nanoparticle size and excitation wavelength, the new and enhanced vibrational modes were ascribed to charge transfer from valence band in PbS QDs to molecule, as shown in Figure 8.

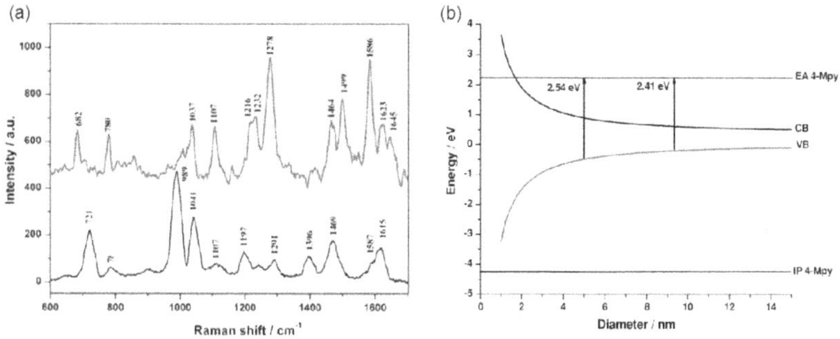

Figure 8. (a) Comparison of the Raman spectrum of 4-Mpy powder (lower) with that of 4-Mpy adsorbed on 8.9 nm PbS quantum dot (upper) under excitation at 514.5 nm. (b) CT from size-dependent VB of PbS QDs to LUMO of 4-Mpy. Reproduced with permission from reference (34). Copyright 2011 American Institute of Physics.

Zhao's group focuses on the CT research of diverse metal-semiconductor systems (*87*, *88*). They investigated the interaction of Cu-ZnO nanorod surfaces with dipolar organic molecule layer of 4-aminothiophenol (PATP) by using variable (UV-visible-infrared) excitation Raman spectroscopy (Figure 9) (*89*). SERS spectra demonstrated the occurrence of chemical enhancement between ZnO nanorods and PATP at visible-infrared excitation, and the observation of additional strong enhancements of multiphonon modes of ZnO nanorods on plasmonic copper sheet excited by a 325 nm laser indicated the surface plasmon resonance effect facilitates CT resonance process between ZnO-molecule.

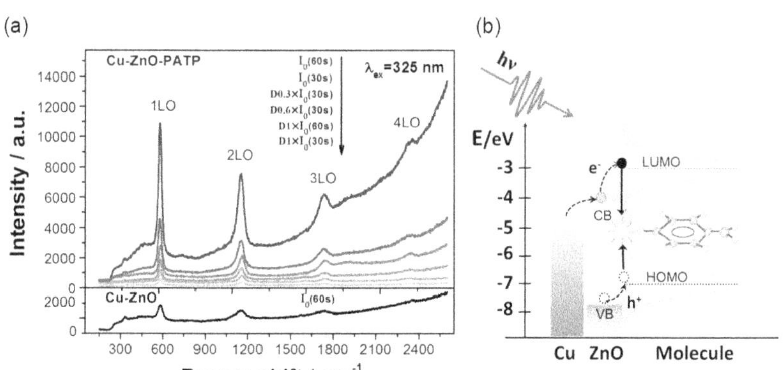

Figure 9. (a) Multiphonon resonant Raman scattering spectra of Cu-ZnO-PATP model (top) excited with powers of I_0 and $D1\times I_0(10^{-1}I_0)$ at time interval of 60 s, I_0, $D0.3\times I0$ $(10^{-0.3}I_0)$, $D0.6\times I_0$ $(10^{-0.6}I_0)$ and $D1\times I_0$ $(10^{-1}I_0)$ at time interval of 30 s, respectively, and Cu-ZnO substrate excited with powers of I_0 at time interval of 60 s (bottom). (b) Schematic energy-level diagram of the HOMO and the LUMO of PATP with respect to the conduction band (CB) and valence band (VB) after contacts. Reproduced with permission from reference (89). Copyright 2012 American Chemical Society.

Besides Zhao's and co-workers' studies, theoretical and experimental research on CT in various metal-semiconductor-molecule systems are still remains insufficient and particularly required to be well investigated, not only for efficient molecule detection in practical biochemistry or forensic science but also for better understanding of optical or electronic properties of the complex system, such as a strong coupled system that metal and semiconductor or molecule form hybridized states and modify optical and vibrational spectrum.

Photochemical Reaction Detected by SERS

SERS is a powerful diagnostic tool and a novel approach to probe the dynamic surface/interface chemistry of quantum dots, especially when they involve in oxidative, catalytic, and electrochemical surface/interface reactions. Chemical reactions of semiconductor QDs have been revealed by SERS. Recently, İlker Doğan and co-workers (*90*) investigated oxidation process of Si QDs on Ag/Ag_2O films and observed an emergence of $Si\text{-}O_x$ and $Si\text{-}O\text{-}H_x$ modes and a gradual decrease of $Si\text{-}Cl_x$ and $Si\text{-}H_x$ modes, which covered the surface of the as-synthesized Si QDs surfaces.

PbS QDs coupled to LSPR has been demonstrated as a promising system for efficient photoenergy conversion owing both to the enhanced electromagnetic field as well as the potential to preserve multiple exciton generation. Size-controlled PbS QDs sensitized $TiO_2/Au/TiO_2$ working electrode has been proved to greatly enhance photoelectric response across a wide wavelength range (*33*), and for the same system, photochemical reaction between excited PbS QDs and sulfide redox couple (S^{2-}/S) in the electrolyte has been revealed by in-situ electrochemical-SERS and photoelectric response measurement (*91*). As shown in Figure10, the formation of vibrational modes at 155, 222 and 474 cm^{-1} at potential of +0.1 V were assigned as polysulfur (S_8) bending and stretching according to a previous studies (*92*, *93*), demonstrating the oxidation of the donor of sulfide (S^{2-}) by the hole generated by photoexcited PbS QDs. The enlarged potential gradient at a positive potential polarization benefit electron-hole separation, so that the electron injected into the conduction band of TiO_2 is collected as photocurrent up to 40 nA while the hole oxides sulfide into polysulfur and characterized as Raman signals. In addition, the enhanced electromagnetic field of coupling PbS QDs with Au NPs not only facilitates the electron injection, but also assist Raman detection. Such LSPR-semiconductor QDs coupling system provide a way to pave reaction processes between electrode surface and electrolyte.

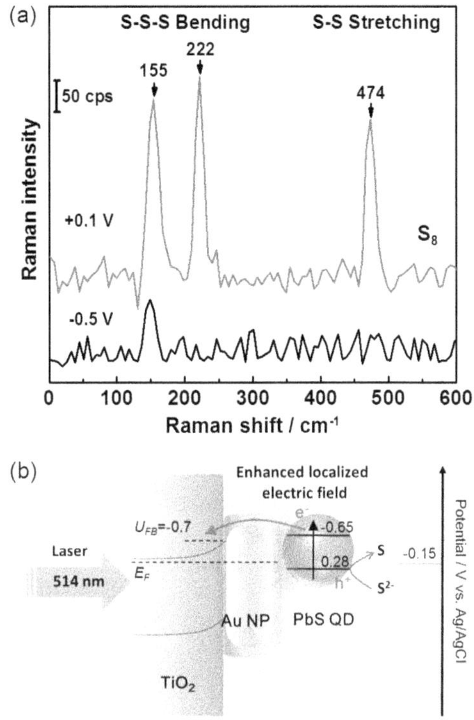

Figure 10. (a) EC-SERS spectrum of PbS QDs/TiO_2/Au/TiO_2 at potential of -0.5 V and +0.1 V (vs. Ag/AgCl) in electrolyte of 0.05 M Na_2S+0.1 M NaOH aqueous solution. (b) Schematic illustration of PbS QDs/TiO_2/Au/TiO_2 electrode in contact with S^{2-}/S redox couple. The conduction and valence band edge position was calculated in regard to the average size evaluated from the 1st exciton wavelength of 1344 nm (94, 95). The flat band energy of TiO_2 with respect to Ag/AgCl (sat.KCl) refer to anatase (96, 97) in alkaline aqueous solution (pH=13). Enhanced near-field generated by LSPR is confined between Au NP and PbS QD (32, 33). Reproduced from reference (91).

Conclusion and Perspective

In this account, we have introduced and explained the enhancement mechanisms of photoexcitation of semiconductor nanostructures-LSPR systems via SERS spectroscopy. Besides of electromagnetic field of LSPR, vibrational spectrum of semiconductor nanostructures could also be influenced by factors such as electron-phonon coupling of QDs in electromagnetic field, plasmon heating, and even the formation of hybridized states between LSPR and semiconductor nanostructures in the strong coupling regime, however, theoretical and experimental studies still remain blank and could be a direction for a better understanding the intrinsic properties of QDs in highly confined electromagnetic field.

Future challenges of in the development of SERS in metal-semiconductor-molecule systems may include investigation of exciton-vibration coupling between semiconductor nanostructures and capping molecules, as vibration spectrum provide by SERS information for understanding exciton recombination processes in semiconductor nanostructures, and promisingly provide insight to better design and apply in for efficient light harvesting, photo emitting, photoelectric or photochemical conversion in an expanded solar spectrum.

References

1. Smith, E.; Dent G. Surface-Enhanced Raman Scattering and Surface-Enhanced Resonance Raman Scattering. In *Modern Raman Spectroscopy – A Practical Approach*; John Wiley & Sons: Chichester, England, 2005; pp 113−133.
2. Kneipp, K.; Kneipp, H.; Itzkan, I.; Dasari, R. R.; Feld, M. S. *Chem. Rev.* **1999**, *99*, 2957–2976.
3. Kneipp, J.; Kneipp, H.; Kneipp, K. *Chem. Soc. Rev.* **2008**, *37*, 1052–1060.
4. Nabiev, I.; Chourpa, I.; Manfait, M. *J. Raman Spectrosc.* **1994**, *25*, 13–23.
5. Lane, L. A.; Qian, X.; Nie, S. *Chem. Rev.* **2015**, *115*, 10489–10529.
6. Livingstone, R.; Zhou, X.; Tamargo, M. C.; Lombardi, J. R.; Quagliano, L. G.; Jean-Mary, F. *J. Phys. Chem. C* **2010**, *114*, 17460–17464.
7. Li, W.; Zamani, R.; Rivera Gil, P.; Pelaz, B.; Ibáñez, M.; Cadavid, D.; Shavel, A.; Alvarez-Puebla, R. A.; Parak, W. J.; Arbiol, J.; Cabot, A. *J. Am. Chem. Soc.* **2013**, *135*, 7098–7101.
8. Cong, S.; Yuan, Y.; Chen, Z.; Hou, J.; Yang, M.; Su, Y.; Zhang, Y.; Li, L.; Li, Q.; Geng, F.; Zhao, Z. *Nat. Commun.* **2015**, *6*.
9. Lombardi, J. R.; Birke, R. L. *Acc. Chem. Res.* **2009**, *42*, 734–742.
10. Tian, Z. Q. *J. Raman Spectrosc.* **2005**, *36*, 466–470.
11. Huang, Y.; Fang, Y.; Zhang, Z.; Zhu, L.; Sun, M. *Light Sci. Appl.* **2014**, *3*, e199.
12. Pettinger, B.; Schambach, P.; Villagómez, C. J.; Scott, N. *Annu. Rev. Phys. Chem.* **2012**, *63*, 379–399.
13. Sonntag, M. D.; Pozzi, E. A.; Jiang, N.; Hersam, M. C.; Van Duyne, R. P. *J. Phys. Chem. Lett.* **2014**, *5*, 3125–3130.
14. Mubeen, S.; Lee, J.; Lee, W.-r.; Singh, N.; Stucky, G. D.; Moskovits, M. *ACS Nano* **2014**, *8*, 6066–6073.
15. Linic, S.; Christopher, P.; Ingram, D. B. *Nat. Mater.* **2011**, *10*, 911–921.
16. Atwater, H. A.; Polman, A. *Nat. Mater.* **2010**, *9*, 205–213.
17. Clavero, C. *Nat. Photonics* **2014**, *8*, 95–103.
18. Lozano, G.; Rodriguez, S. R. K.; Verschuuren, M. A.; Gomez Rivas, J. *Light Sci. Appl.* **2016**, *5*, e16080.
19. Jin, Y.; Gao, X. *Nat. Nanotechnol.* **2009**, *4*, 571–576.
20. Meixner, A. J.; Jager, R.; Jager, S.; Brauer, A.; Scherzinger, K.; Fulmes, J.; Krockhaus, S. z. O.; Gollmer, D. A.; Kern, D. P.; Fleischer, M. *Faraday Discuss.* **2015**, *184*, 321–337.
21. Mayer, K. M.; Hafner, J. H. *Chem. Rev.* **2011**, *111*, 3828–3857.

22. Anker, J. N.; Hall, W. P.; Lyandres, O.; Shah, N. C.; Zhao, J.; Van Duyne, R. P. *Nat. Mater.* **2008**, *7*, 442–453.
23. Li, M.; Cushing, S. K.; Wu, N. *Analyst* **2015**, *140*, 386–406.
24. Jain, P. K.; Ghosh, D.; Baer, R.; Rabani, E.; Alivisatos, A. P. *Proc. Natl. Acad. Sci. U.S.A.* **2012**, *109*, 8016–8019.
25. Takase, M.; Ajiki, H.; Mizumoto, Y.; Komeda, K.; Nara, M.; Nabika, H.; Yasuda, S.; Ishihara, H.; Murakoshi, K. *Nat. Photonics* **2013**, *7*, 550–554.
26. Nootz, G.; Padilha, L. A.; Levina, L.; Sukhovatkin, V.; Webster, S.; Brzozowski, L.; Sargent, E. H.; Hagan, D. J.; Van Stryland, E. W. *Phys. Rev. B* **2011**, *83*, 155302.
27. Guyot-Sionnest, P.; Wehrenberg, B.; Yu, D. *J. Chem. Phys* **2005**, *123*, 074709.
28. Lifshitz, E. *J. Phys. Chem. Lett.* **2015**, *6*, 4336–4347.
29. Peterson, M. D.; Cass, L. C.; Harris, R. D.; Edme, K.; Sung, K.; Weiss, E. A. *Annu. Rev. Phys. Chem.* **2014**, *65*, 317–339.
30. Nelson, C. A.; Monahan, N. R.; Zhu, X. Y. *Energy Environ. Sci.* **2013**, *6*, 3508–3519.
31. Sambur, J. B.; Novet, T.; Parkinson, B. A. *Science* **2010**, *330*, 63–66.
32. Hutter, T.; Mahajan, S.; Elliott, S. R. *J. Raman Spectrosc.* **2013**, *44*, 1292–1298.
33. Li, X.; Suzuki, K.; Toda, T.; Yasuda, S.; Murakoshi, K. *J. Phys. Chem. C* **2015**, *119*, 22092–22101.
34. Fu, X.; Pan, Y.; Wang, X.; Lombardi, J. R. *J. Chem. Phys* **2011**, *134*, 024707.
35. Kittel, C. Phonons I. Crystal Vibrations. In *Introduction to Solid State Physics*, 8th ed.; John Wiley and Sons Inc.: New York, 2005; pp 89−104.
36. Arora, A. K.; Rajalakshmi, M.; Ravindran, T. R.; Sivasubramanian, V. *J. Raman Spectrosc.* **2007**, *38*, 604–617.
37. Belogorokhov, A. I.; Gavrilov, S. A.; Belogorokhov, I. A.; Tikhomirov, A. A. *Semiconductors* **2005**, *39*, 243–248.
38. Madelung, O. Cadmium Selenide (CdSe) Phonon Wavenumbers, Mean Square Displacements. In *II-VI and I-VII Compounds; Semimagnetic Compounds*; Rössler, U., Schulz, M., Eds.; Springer: Berlin, 1999; Vol. 41B, pp 1−4.
39. Nanda, K. K.; Sahu, S. N.; Soni, R. K.; Tripathy, S. *Phys. Rev. B* **1998**, *58*, 15405–15407.
40. Beard, M. C.; Ellingson, R. J. *Laser Photon. Rev.* **2008**, *2*, 377–399.
41. Harbold, J. M.; Du, H.; Krauss, T. D.; Cho, K.-S.; Murray, C. B.; Wise, F. W. *Phys. Rev. B* **2005**, *72*, 195312.
42. Geiregat, P.; Delerue, C.; Justo, Y.; Aerts, M.; Spoor, F.; Van Thourhout, D.; Siebbeles, L. D. A.; Allan, G.; Houtepen, A. J.; Hens, Z. *ACS Nano* **2015**, *9*, 778–788.
43. Blackburn, J. L.; Chappell, H.; Luther, J. M.; Nozik, A. J.; Johnson, J. C. *J. Phys. Chem. Lett.* **2011**, *2*, 599–603.
44. Prabhu, R. R.; Abdul Khadar, M. *Bull. Mater. Sci* **2008**, *31*, 511–515.
45. Ketterer, B.; Heiss, M.; Uccelli, E.; Arbiol, J.; Fontcuberta i Morral, A. *ACS Nano* **2011**, *5*, 7585–7592.

46. Cheng, H.-M.; Lin, K.-F.; Hsu, H.-C.; Hsieh, W.-F. *Appl. Phys. Lett.* **2006**, *88*, 261909.
47. Lemos, V.; Silveira, E.; Leite, J. R.; Tabata, A.; Trentin, R.; Scolfaro, L. M. R.; Frey, T.; As, D. J.; Schikora, D.; Lischka, K. *Phys. Rev. Lett.* **2000**, *84*, 3666–3669.
48. Huang, K.; Rhys, A. *Proc. R. Soc. London, Ser. A* **1950**, *204*, 406–423.
49. Krauss, T. D.; Wise, F. W.; Tanner, D. B. *Phys. Rev. Lett.* **1996**, *76*, 1376–1379.
50. Pokatilov, E. P.; Klimin, S. N.; Fomin, V. M.; Devreese, J. T.; Wise, F. W. *Phys. Rev. B* **2002**, *65*, 075316.
51. Schaller, R. D.; Pietryga, J. M.; Goupalov, S. V.; Petruska, M. A.; Ivanov, S. A.; Klimov, V. I. *Phys. Rev. Lett.* **2005**, *95*, 196401.
52. Minnaert, A. W. E.; Silov, A. Y.; van der Vleuten, W.; Haverkort, J. E. M.; Wolter, J. H. *Phys. Rev. B* **2001**, *63*, 075303.
53. Khajehpour, J.; Daoud, W. A.; Williams, T.; Bourgeois, L. *J. Phys. Chem. C* **2011**, *115*, 22131–22137.
54. Sahoo, S.; Arora, A. K. *J. Phys. Chem. B* **2010**, *114*, 4199–4203.
55. Freitas Neto, E. S.; da Silva, S. W.; Morais, P. C.; Dantas, N. O. *J. Phys. Chem. C* **2013**, *117*, 657–662.
56. Dzhagan, V. M.; Valakh, M. Y.; Himcinschi, C.; Milekhin, A. G.; Solonenko, D.; Yeryukov, N. A.; Raevskaya, O. E.; Stroyuk, O. L.; Zahn, D. R. T. *J. Phys. Chem. C* **2014**, *118*, 19492–19497.
57. Cherevkov, S. A.; Fedorov, A. V.; Artemyev, M. V.; Prudnikau, A. V.; Baranov, A. V. *Phys. Rev. B* **2013**, *88*, 041303.
58. Krauss, T. D.; Wise, F. W. *Phys. Rev. B* **1997**, *55*, 9860–9865.
59. Kelley, A. M. *J. Phys. Chem. Lett.* **2010**, *1*, 1296–1300.
60. Darugar, Q.; Landes, C.; Link, S.; Schill, A.; El-Sayed, M. A. *Chem. Phys. Lett.* **2003**, *373*, 284–291.
61. Kilina, S. V.; Kilin, D. S.; Prezhdo, O. V. *ACS Nano* **2009**, *3*, 93–99.
62. Semonin, O. E.; Luther, J. M.; Choi, S.; Chen, H.-Y.; Gao, J.; Nozik, A. J.; Beard, M. C. *Science* **2011**, *334*, 1530–1533.
63. Wang, L.-W.; Califano, M.; Zunger, A.; Franceschetti, A. *Phys. Rev. Lett.* **2003**, *91*, 056404.
64. Cho, B.; Peters, W. K.; Hill, R. J.; Courtney, T. L.; Jonas, D. M. *Nano Lett.* **2010**, *10*, 2498–2505.
65. Trinh, M. T.; Sfeir, M. Y.; Choi, J. J.; Owen, J. S.; Zhu, X. *Nano Lett.* **2013**, *13*, 6091–6097.
66. Schatz, G. C.; Young, M. A.; Van Duyne, R. P. Electromagnetic Mechanism of SERS. In *Surface-Enhanced Raman Scattering: Physics and Applications*; Kneipp, K., Moskovits, M., Kneipp, H., Eds.; Springer-Verlag: New York, 2006; pp 19–45.
67. Ronald L. Birke, J. R. L. Surface-Enhanced Raman Scattering. In *Spectroelectrochemistry: Theory and Practice*; Gale, R. J., Ed.; Springer US: Plenum Press: New York, 1998; pp 263–348.
68. Hakami, J.; Wang, L.; Zubairy, M. S. *Phys. Rev. A* **2014**, *89*, 053835.
69. Wang, T.; Zhang, Z.; Liao, F.; Cai, Q.; Li, Y.; Lee, S.-T.; Shao, M. *Sci. Rep.* **2014**, *4*, 4052.

70. Moreels, I.; Kruschke, D.; Glas, P.; Tomm, J. W. *Opt. Mater. Express* **2012**, *2*, 496–500.
71. Tanya, H.; Stephen, R. E.; Sumeet, M. *Nanotechnology* **2013**, *24*, 035201.
72. Milekhin, A. G.; Yeryukov, N. A.; Sveshnikova, L. L.; Duda, T. A.; Rodyakina, E. E.; Sheremet, E. S.; Ludemann, M.; Gordan, O. D.; Latyshev, A. V.; Zahn, D. R. T. *Thin Solid Films* **2013**, *543*, 35–40.
73. Taz, H.; Ruther, R.; Malasi, A.; Yadavali, S.; Carr, C.; Nanda, J.; Kalyanaraman, R. *J. Phys. Chem. C* **2015**, *119*, 5033–5039.
74. Degioanni, S.; Jurdyc, A. M.; Cheap, A.; Champagnon, B.; Bessueille, F.; Coulm, J.; Bois, L.; Vouagner, D. *J. Appl. Phys.* **2015**, *118*, 153103.
75. Rumyantseva, A.; Kostcheev, S.; Adam, P.-M.; Gaponenko, S. V.; Vaschenko, S. V.; Kulakovich, O. S.; Ramanenka, A. A.; Guzatov, D. V.; Korbutyak, D.; Dzhagan, V.; Stroyuk, A.; Shvalagin, V. *ACS Nano* **2013**, *7*, 3420–3426.
76. Hugall, J. T.; Baumberg, J. J.; Mahajan, S. *Appl. Phys. Lett.* **2009**, *95*, 141111.
77. Sigle, D. O.; Hugall, J. T.; Ithurria, S.; Dubertret, B.; Baumberg, J. J. *Phys. Rev. Lett.* **2014**, *113*, 087402.
78. Todescato, F.; Minotto, A.; Signorini, R.; Jasieniak, J. J.; Bozio, R. *ACS Nano* **2013**, *7*, 6649–6657.
79. Wang, X.; Shi, W.; She, G.; Mu, L. *Phys. Chem. Chem. Phys.* **2012**, *14*, 5891–5901.
80. Lombardi, J. R.; Birke, R. L. *J. Phys. Chem. C* **2014**, *118*, 11120–11130.
81. Quagliano, L. G. *J. Am. Chem. Soc.* **2004**, *126*, 7393–7398.
82. Livingstone, R.; Zhou, X.; Tamargo, M. C.; Lombardi, J. R.; Quagliano, L. G.; Jean-Mary, F. *J. Phys. Chem. C* **2010**, *114*, 17460–17464.
83. Musumeci, A.; Gosztola, D.; Schiller, T.; Dimitrijevic, N. M.; Mujica, V.; Martin, D.; Rajh, T. *J. Am. Chem. Soc.* **2009**, *131*, 6040–6041.
84. Xue, X.; Ji, W.; Mao, Z.; Mao, H.; Wang, Y.; Wang, X.; Ruan, W.; Zhao, B.; Lombardi, J. R. *J. Phys. Chem. C* **2012**, *116*, 8792–8797.
85. Islam, S. K.; Tamargo, M.; Moug, R.; Lombardi, J. R. *J. Phys. Chem. C* **2013**, *117*, 23372–23377.
86. Li, W.; Zamani, R.; Gil, P. R.; Pelaz, B.; Ibanez, M.; Cadavid, D.; Shavel, A.; Alvarez-Puebla, R. A. W.; Parak, J.; Arbiol, J.; Cabot, A. *J. Am. Chem. Soc.* **2013**, *135*, 7098–7101.
87. Zhang, X.; Yu, Z.; Ji, W.; Sui, H.; Cong, Q.; Wang, X.; Zhao, B. *J. Phys. Chem. C* **2015**, *119*, 22439–22444.
88. Mao, Z.; Song, W.; Chen, L.; Ji, W.; Xue, X.; Ruan, W.; Li, Z.; Mao, H.; Ma, S.; Lombardi, J. R.; Zhao, B. *J. Phys. Chem. C* **2011**, *115*, 18378–18383.
89. Mao, Z.; Song, W.; Xue, X.; Ji, W.; Chen, L.; Lombardi, J. R.; Zhao, B. *J. Phys. Chem. C* **2012**, *116*, 26908–26918.
90. Doğan, İ.; Gresback, R.; Nozaki, T.; van de Sanden, M. C. M. *Sci. Rep.* **2016**, *6*, 29508.
91. Li, X.; Suzuki, K.; Yoshii, T.; Yasuda, S.; Murakoshi, K. *Surface-Enhanced Raman Scattering Observation by Coupled PbS Quantum Dots-Au Nanoparticle System*, the international chemical congress of pacific basin societies (Pacifichem), Hawaii, USA, 2015, PHYS 869.

92. Hagen, M.; Schiffels, P.; Hammer, M.; Dörfler, S.; Tübke, J.; Hoffmann, M. J.; Althues, H.; Kaskel, S. *J. Electrochem. Soc.* **2013**, *160*, A1205–A1214.
93. Parker, G. K.; Watling, K. M.; Hope, G. A.; Woods, R. *Colloids Surf., A* **2008**, *318*, 151–159.
94. Moreels, I.; Lambert, K.; Smeets, D.; De Muynck, D.; Nollet, T.; Martins, J. C.; Vanhaecke, F.; Vantomme, A.; Delerue, C.; Allan, G.; Hens, Z. *ACS Nano* **2009**, *3*, 3023–3030.
95. Jasieniak, J.; Califano, M.; Watkins, S. E. *ACS Nano* **2011**, *5*, 5888–5902.
96. Kavan, L.; Grätzel, M.; Gilbert, S. E.; Klemenz, C.; Scheel, H. J. *J. Am. Chem. Soc.* **1996**, *118*, 6716–6723.
97. Beranek, R. *Adv. Phys. Chem.* **2011**, *2011*, 20.

Chapter 9

Near-Infrared Excited Surface-Enhanced Raman and Hyper Raman Scattering for Microscopic Mapping of Biosamples

Janina Kneipp,[*,1] Marina Gühlke,[1] and Zsuzsanna Heiner[1,2]

[1]Humboldt-Universität zu Berlin, Department of Chemistry, Brook-Taylor-Str. 2, 12589 Berlin, Germany
[2]Humboldt-Universität zu Berlin, School of Analytcial Sciences Adlershof, Albert-Einstein-Str. 5-9, 12489 Berlin, Germany
[*]E-mail: janina.kneipp@chemie.hu-berlin.de

The chapter will discuss spontaneous, non-resonant and resonant plasmon-enhanced Raman scattering for the characterization of microscopic samples based on multi-photon excitation. Surface-enhanced hyper Raman scattering (SEHRS), the two-photon analogue of surface-enhanced Raman scattering (SERS) can be used to characterize cells and other microscopic objects based on vibrational information of reporter molecules. A combination of SERS and SEHRS is particularly interesting for the construction of nanosensors that can be used in complex biological environments to measure pH. The combination of SERS and SEHRS spectra excited at 1064 nm allows more robust pH sensing than SERS measurements in the visible or NIR excitation range alone. As another aspect, hyperspectral mapping of microscopically heterogeneous samples using SEHRS spectra is demonstrated. Combined SEHRS /SERS hyperspectral imaging by multivariate analysis can be developed into a useful imaging modality for microscopic biological samples such as cells and tissues.

Advantages of Surface-Enhanced Hyper Raman Scattering for Bioprobing

Two-photon excitation is gaining rapidly in interest and significance in the microspectroscopic characterization of cells and tissues. The main reasons for this are the greatly reduced possible phototoxicity and stress to the sample due to longer excitation wavelengths, and the confinement of the two-photon interaction to the focus of the laser beam. In this chapter, two-photon excited surface-enhanced Raman scattering (SERS), namely surface-enhanced hyper-Raman scattering (SEHRS) will be discussed in the context of spatially resolved vibrational analysis.

Hyper Raman scattering (HRS) is a two-photon excited Raman scattering process, and thus results in Raman signals shifted relative to the doubled energy of the excitation laser (*1*, *2*), a schematic is displayed in Figure 1. HRS follows symmetry selection rules different from regular one-photon Raman scattering. It can probe infrared (IR)-active vibrations that are usually not evident in Raman spectra and in addition reveals so–called "silent" modes, vibrations that are seen neither in Raman nor in IR absorption spectra. For example, the Raman-active vibrational modes of a centrosymmetric molecule are hyper-Raman-forbidden, and those inactive in both IR and Raman can be active in hyper-Raman scattering (*3*). Therefore, the spectral information obtained in HRS is complementary to the information content of other vibrational (one-photon Raman and IR) spectra and could be of use for a number of applications in biophysics and bioanalytical chemistry.

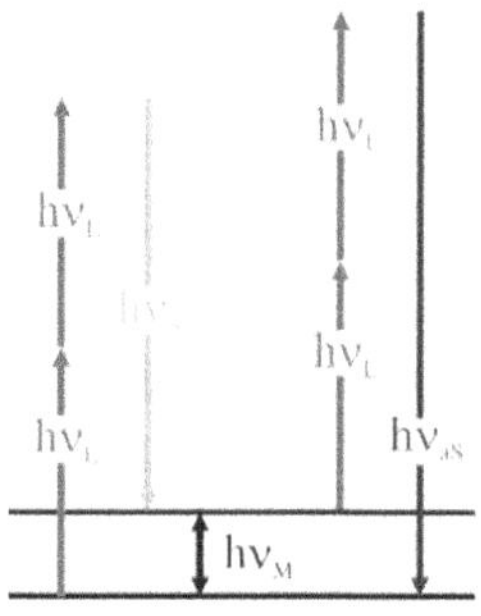

Figure 1. Schematic of Stokes and anti-Stokes hyper Raman scattering.

As a nonlinear, incoherent Raman process, HRS is extremely weak, with scattering cross sections on the order of 10^{-65} $cm^4 s$ $photon^{-1}$, 35 orders of magnitude smaller than cross sections of one-photon-excited Raman scattering and ~15 orders of magnitude below typical two-photon absorption cross sections. These extremely small cross sections have precluded application of HRS as practical spectroscopic tool for a long time. However, as we and others have shown, due to the non-linear dependence on the excitation field, HRS can benefit much more from the high local optical fields than normal Raman scattering does in the case of SERS (*4*). Assuming effective SERS cross sections on the order of 10^{-16} cm^2 (*5–7*), effective SEHRS cross sections were found to be on the order of 10^{-46} cm^4s, and correspond to 10^4 GM (Goeppert-Mayer) (*4*), similar or even higher than typical cross sections in two-photon fluorescence (*8*, *9*).

Compared with its SERS spectrum, the SEHRS spectrum of a centrosymmetric molecule can contain new vibrational bands and display significant relative intensity differences because of the surface effect (*10*). Different from this, the two-photon excited spectrum for a non-centrosymmetric molecule largely resembles its SERS spectrum (*4*, *11*, *12*). Figure 2 gives examples of SEHRS spectra from cells. They show, e.g., pronounced amide II signals of proteins, usually detected by IR absorption spectroscopy (around 1540 cm^{-1}, please see the IR spectrum obtained from the cytoplasmic region of the same cell for comparison). Interestingly, SEHRS observes also spectral bands above 1800 cm^{-1}, which are likely combination modes. The spectral region between 1800 and 2700 cm^{-1} is almost featureless in other vibrational (Raman and IR) spectra of cells and tissues.

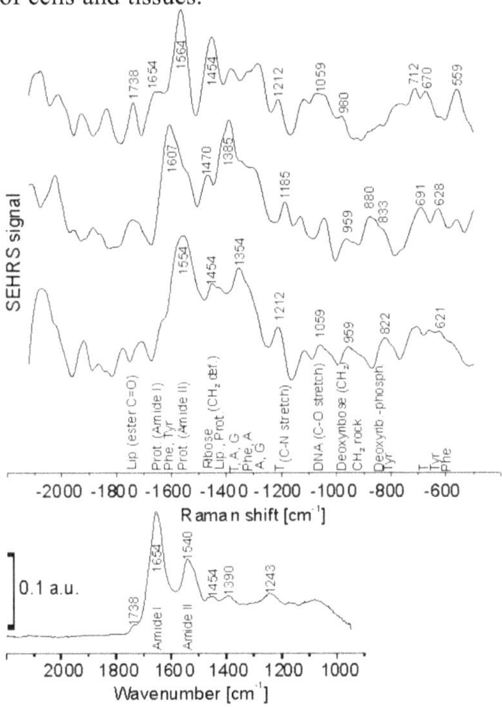

Figure 2. SEHRS spectra measured from J 774 cells after 4 hours uptake of gold nanoparticles. Excitation was achieved with 1064 nm mode locked ps pulses, average power 30 mW, size of the laser spot ~1 μm, collection time 10 seconds. Example of an IR spectrum of a dried J 774 cell on CaF_2 substrate. The IR spectrum was acquired from a ~30 μm-diameter spot. Reproduced with permission from (4). Copyright (2006) National Academy of Sciences, U.S.A.

The strong confinement of the excitation of HRS to the focus of the laser poses several advantages for microscopic probing, as well as several challenges for the experimentalist. From typical experiments it was estimated that the focal volume in a SEHRS experiment was ~30 μm³, significantly smaller than in a typical SERS experiment (*13*). This high localization and small focal volumes enable SEHRS microscopic probing at high resolution and of very few nanoparticles and molecules.

Using laser light in the NIR wavelength range in order to excite SEHRS, is also interesting to analyze the SERS spectra of the same sample. Until recently, in the excitation range between 1000 nm and 1300 nm, mostly Fourier-transform (FT) techniques have been used in Raman spectroscopy, but in the past few years technological advances allowed a more widespread use of dispersive set-ups (*14*, *15*). SERS was observed for NIR excitation first in FT-Raman on electrodes (*16–18*) and later also on silver nanoparticles (*19*, *20*). Ideally, both NIR-excited processes (NIR-SERS and SEHRS) combined together will give different insights into molecular symmetry and in a comprehensive way could be used to investigate different molecules and molecule-nanostructure interactions. This aspect will be discussed in this chapter as well. As will be illustrated in the following, two-photon excited SERS for studies of biological systems may provide new molecular information, combined with the advantages of both excitation in the near-infrared and the sensitivity and improved lateral resolution of plasmonics-based spectroscopy.

pH Probing in Animal Cells with SERS Reporter Molecules

Probing of chemical parameters at the microscopic level by SERS can be achieved in two ways. (i) through direct probing of composition, structure, and interaction of biomolecules with one another and / or with the nanostructure (*21–23*) and (ii) by using of a reporter molecule with a spectrum that can indicate changes in local chemical parameters.

pH is an important microenvironmental parameter, e.g., in biological samples such as cells and tissues, or in microfluidic systems, and the SERS spectrum of several molecules was shown to depend on pH. *para*-mercaptobenzoic acid (*p*MBA), similar to other organothiols (*24*), can strongly attach to silver nanostructures via its thiol group (*25*, *26*). Concentration-dependent changes of the orientation of the molecules on the silver surface were observed in SERS spectra (*27*). Moreover, in the spectrum of *p*MBA, the protonation and deprotonation of the carboxylate group in different pH environment can be observed (*27*), therefore the *p*MBA molecule can be used as a SERS pH nanosensor (*26*, *28*, *29*). The pH dependent SERS spectra of *p*MBA adsorbed on gold nanoshells bound to a silicon substrate were used to create a pH meter working over the range of 5.8 to 7.6 pH units (*30*). Similar experiments demonstrated that hollow gold nanospheres with *p*MBA were responsive over a pH range of 3.5 to 9 (*31*). In many cell types, significant acidification takes place when endosomal structures mature to become late endosomes and lysosomes (*32*). In the lysosome, to ensure e.g., proper enzymatic function, pH can be well below 5, and even below 4 (*33*). From SERS spectra measured with silver particles and *p*MBA in ovary cells it was concluded that the pH in the environment of the particles was below 6, an observation consistent with the particles being located inside lysosomes (*26*). Spectra of *p*MBA on gold nanospheres indicated variation of pH in the endosomal system between 6.8 and 5.4 in endosomes of different ages (*29*).

In these experiments, endosomes of different age were obtained using different incubation times with the nanosensors. Individual cells were raster-scanned in a Raman microspectroscopic setup. The pH in the different sampling volumes was determined based on the ratio of two characteristic bands in the spectrum of *p*MBA, one with varying and one with constant intensity, respectively. The color map in Figure 3 displays the ratio of the SERS signals at 1423 cm^{-1} and 1076 cm^{-1} as a function of sampling position in one of the cells. Analyzing the spectra from a whole population of cells at different time points, it is possible to observe an increase in sampled spots with low pH. Although each sampled spot can contain several endosomal structures, this indicates that there is an overall increase in late endosomal and lysosomal structures (Figure 3), with a significant portion of very acidic volumes after 6 hours of incubation of the cells with the nanoprobes.

Considering that pH in lysosomes is significantly lower than 5.4, a pH nanosensor used in this way does not provide the full information about endosomal pH. The pH range below 5.4 cannot be further resolved, since the SERS signals that were used were not sensitive to changes in this acidic range. In the following, increasing the sensitivity of *p*MBA pH sensors by using two-photon excitation will be discussed.

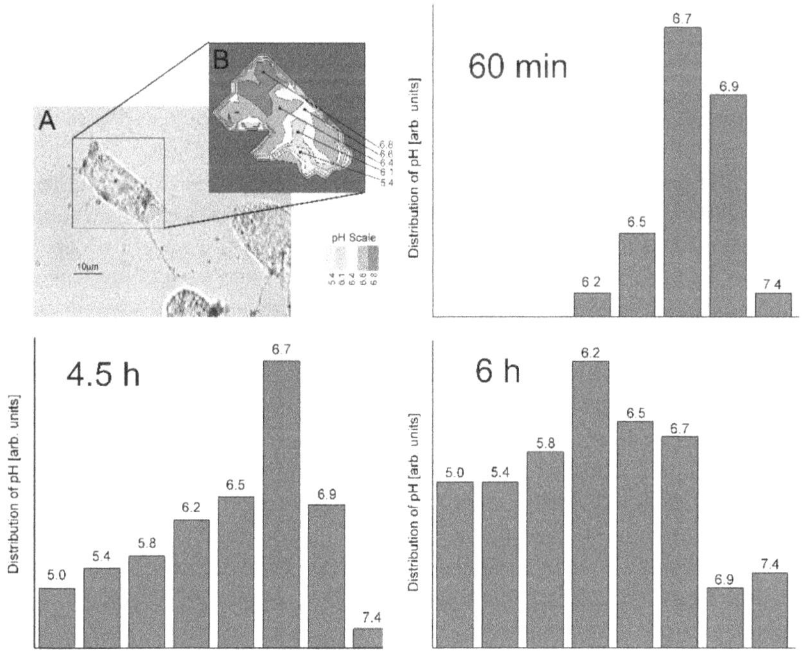

Figure 3. Mapping of pH using the signals of the spectrum of pMBA inside a cell. Distribution of pH values after different incubation times with the pH probes. Reproduced with permission from (29). Copyright (2010, ACS).

SEHRS and NIR-SERS Spectra of pH Reporter *p*MBA

SEHRS spectra of *p*MBA are very similar to SERS spectra, but the relative signal strengths of the bands in the spectrm are changed. In a first experiment, pH sensors consisting of *p*MBA on silver nanoparticles showed that the band assigned to the COO^- vibration in the deprotonated molecule at 1380 cm^{-1}, which can be used for pH measurements by SERS, appears at higher signal level in the SEHRS and could therefore be followed down to lower pH values. Additionally, the 1700 cm^{-1} COOH stretching mode appeared as pronounced strong band and showed an opposite dependence on pH Based on these two signals with opposite pH dependence, first SEHRS spectra from cells identified pH values over a wide range, from pH 2 to pH 8 (*28*).

Recently, a better understanding of the two-photon excited SEHRS spectrum pf *p*MBA could be obtained. In this chapter, SEHRS spectra of *p*MBA acquired between pH 2 and 12 under non-resonant conditions, together with the visible and near-infrared (1064 nm) excited SERS spectra will be discussed.

Figure 4 shows SEHRS data obtained with citrate stabilized silver nanoparticles (Figure 4a) and with hydroxylamine-stabilized silver nanoparticles (Figure 4b), they are very similar to spectra obtained from live cells that had been reported earlier (*28*). The small difference in some relative band intensities in Figure 4b compared to the spectra in Figure 4a indicate a different interaction of the *p*MBA molecules with the citrate and the hydroxylamine-stabilized silver nanoparticles, respectively. When pH increases, the relative band intensities of vibrations associated with the carboxyl group (marked in green in Figure 4) change relative to those of the aromatic ring (marked in blue in Figure 4). Upon deprotonation, the intensity of the COO^- band at 1365 cm^{-1} increases, and the C=O stretching vibration at 1685 cm^{-1} decreases. This is in accord with the pH dependence of one-photon SERS spectra excited at 532 nm (*34*).

We find in the SEHRS spectra smaller Raman shifts by 5-15 cm^{-1} for several bands, when comparing them to one-photon excited SERS spectra (Table 1). Investigating four different types of silver nanoparticles, it was noticed that these shifts are not specific for one nanoparticle type (*34*). It is very likely, that these bands indicate probing of different vibrations in SERS and in SEHRS due to the different selection rules (*10*), and that the shifts are in fact indicative of probing of different adsorption species. For other molecules it is known that SEHRS is very sensitive to surface potential (*11*) and to local surface environmental changes (*10*), more than SERS.

For high pH values, new band components appear, such as asymmetric broadening and new shoulders, supporting the conclusion that different adsorption species are probed in SERS and SEHRS. Figure 5a displays this for the example of the ring stretching vibration at 1585 cm^{-1} in the SEHRS spectra, while the band in the SERS data is only shifted and shows no second component (Figure 5b,c). The appearance of a low-frequency shoulder in the SEHRS spectrum can be assigned to the non-totally symmetric ring stretching vibration, whereas the high-frequency component corresponds to the totally symmetric ring stretching vibration. The former was reported in SERS data as well, upon increased charge transfer between the molecule and the silver surface (*38*), or when intermolecular interaction between the phenyl ring and a carboxylate group takes place (*40*). From this information we can conclude that at high pH a larger fraction of the molecules must be adsorbed on the surface in a flat orientation and can be specifically probed by SEHRS (Figure 4a). Apart from a small shift this is not visible from the ring stretching band in the SERS spectra (Figure 5b,c). Nevertheless, other bands in the SERS spectra support the same pH-dependent change in orientation as well. The flat orientation is also evidenced by a higher intensity of bands from out-of-plane vibrations of the phenyl ring at 684 cm^{-1} and 710 cm^{-1} in both the SEHRS and SERS spectra at neutral and basic pH (*27*, *35*, *38*).

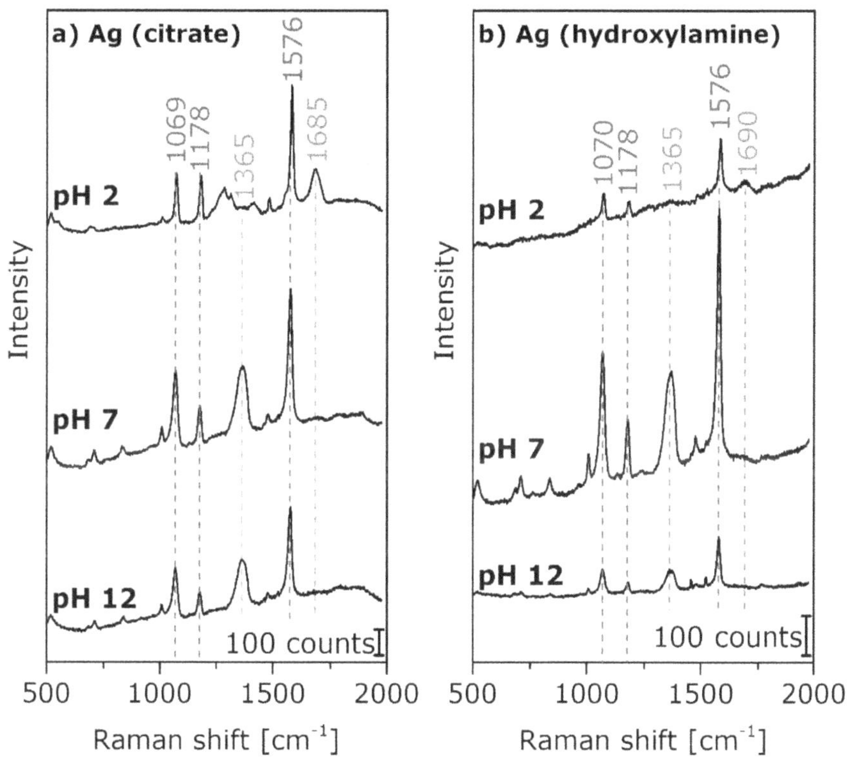

Figure 4. Surface-enhanced hyper Raman spetcra of pMBA in the local field of (a) citrate-stabilized and (b) hydroxylamine-reduced silver nanostructures. Excitation 1064 nm, photon flux density: $2x10^{25}$ photons cm^{-2} s^{-1}, acquisition time 2 s, pMBA concentration $9x10^{-7}$ M, averages of 30 spectra. Reproduced with permission from (34). (Copyright 2015, Published by the PCCP Owner Societies).

As proposed by us and others, pH sensing using *p*MBA SERS / SEHRS spectra relies on changes in relative intensities (*28–30*) that are caused by both protonation and deprotonation of the molecule as well as by the resulting change in orientation at the nanoparticle surface (*27*). The band ratio of the bands at

1365 cm^{-1} (COO^- stretching) and at 1069 cm^{-1} (ring breathing) from the SEHRS spectra changes as a function of pH. Using this band ratio, acidic pH values can be clearly distinguished, supporting the use of the nanosensor in endosomal sensing, or application in other acidic environments. Furthermore, in an application in biological systems, the advantages of two-photon excitation regarding material damage, spatial resolution, and penetration depth are obvious, therefore, using excitation at an infrared wavelength also for one-photon excitation is desirable.

Table 1. Band positions (cm^{-1}) in SEHRS and SERS spectra of *p*MBA with citrate-stabilized silver nanoparticles and band assignments at an excitation wavelength of 1064 nm at pH7. Data originally published in ref. (*34*)

SEHRS (1064 nm)	*SERS (1064 nm)*	*Assignment*	*References*
	363	phenyl deformation + C-S-stretching	(*35, 36*)
520	523	in-plane ring deformation	(*36, 37*)
684		C-H out-of-plane deformation	(*38*)
695*		C-H out-of-plane deformation + out-of-plane γ(CCC)	
710	716	out-of-plane γ(CCC)	(*27, 35, 38*)
836	839	COO^- deformation	(*27, 35, 39*)
1009	1012	ring deformation	(*40*)
1069	1075	ring breathing	(*40*)
	1138	C-COO- stretching	(*40*)
1178	1182	C-H-in-plane-bending	(*40*)
1365		COO^- stretching (COO^- surface-bound)	(*27, 41*)
	1375	COO^- stretching (COO^- non-surface-bound)	(*27*)
1479		C-H in-plane-bending	(*40*)
1576	1583	ring stretching	(*40*)
1685*		C=O stretching of protonated carboxyl group	(*27*)

* Band only observed at acidic pH.

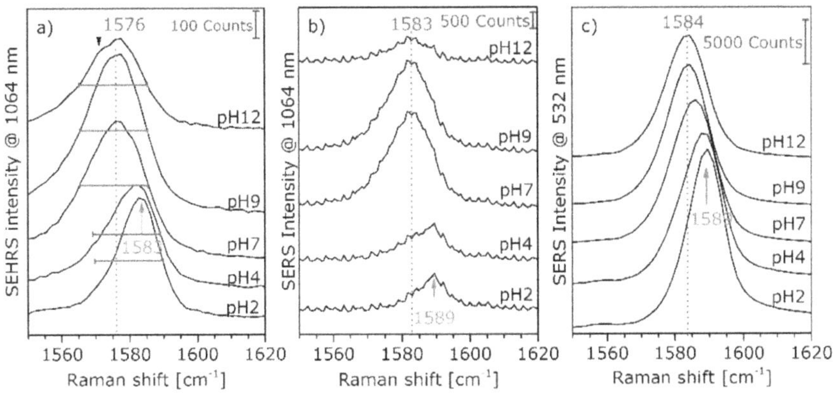

Figure 5. Extracts of (a) SEHRS and (b, c) SERS spectra of pMBA with Ag (citrate) nanoparticles at an excitation wavelength of 1064 nm (b) and 532 nm (c). Reproduced with permission from (13) (Copyright 2015, Published by the PCCP Owner Societies).

In Figure 6, SERS spectra that were excited with 1064 nm by the same laser as the SEHRS spectra are shown. They were detected quasi-simultaneously in one microspectroscopic setup. In the NIR-SERS spectra, the intensity ratio of the pH-sensitive band at 363 cm^{-1} and of the pH-insensitive band at 523 cm^{-1} of a phenyl deformation vibration can be used to discriminate pH values in the range between 2 and 7 (Figure 6b). The band at 363 cm^{-1} can be assigned to deprotonated *p*MBA (*42*), particularly to a phenyl deformation combined with the C-S-stretching vibration (*35, 36*) and increases in intensity with pH becoming less acidic (Figure 6a). Since its intensity is much stronger than that of the band at 710 cm^{-1} of the out-of-plane phenyl γ(C C C), that is used in the SEHRS spectra to probe molecular orientation, it can serve as a very sensitive indicator for pH-induced changes in molecular orientation as well.

The spectra measured with the hydroxylamine-stabilized nanoparticles **in** Figure 4b show a reduction in SEHRS enhancement by a factor of 2-3 at very low and very high pH. Looking at the one-photon excited SERS spectra in Figure 6, a similar observation is made, but the effect was not observed in 532nm excited SERS spectra (*34*) with the same samples. Especially for the acidic pH values, the change in overall enhancement supports a strong influence of an altered interaction of the molecule with the metal nanoparticle surface, as was proposed previously (*27*), or with the stabilizing species (*43*). Both, the stabilizing citrate and hydroxylamine/hydroxide contain functional groups that can be protonated or deprotonated, depending on the surrounding pH. This leads to changes in surface charge and possible aggregation of the nanosensors. For example, a negative surface charge of the silver particles decreases with decreasing pH by protonation, leading to aggregation and minor changes in gap sizes in the nanoaggregates and therefore to a lower electromagnetic enhancement at very acidic pH (*44, 45*). As a further explanation, protonation of the thiolate could weaken the Ag-S-bond and thus change the interaction between *p*MBA itself and the silver surface at acidic pH. On the opposite side, at basic pH, where both, the stabilizing molecules and *p*MBA, are deprotonated, the repulsion between the negative charges of the analyte molecule and the nanoparticle surface is high (*44*), which alters the interaction between the metal surface and *p*MBA and thus leads to decreased enhancement.

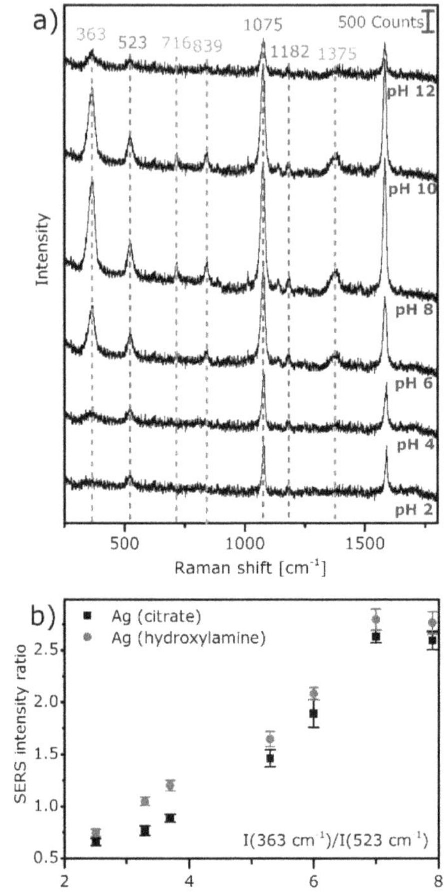

Figure 6. (a) One-photon excited surface enhanced Raman spectra of pMBA with citrate reduced/stabilized AgNP with 1064-nm picosecond laser excitation in different pH environment. Photon flux density: 6 x 10^{25} photons cm^{-2} s^{-1}, acquisition time: 10 s, pMBA concentration: 9 x 10^{-7} M, averages of 30 spectra. (b) Intensity ratios in the spectra of pMBA with citrate (squares) and hydroxylamine (circles) reduced NPs as a function of pH for the bands at 363 and 523 cm^{-1}. Intensity ratios are averaged over 30 spectra, error bars represent the corresponding standard deviations. Reproduced with permission from (13) (Copyright 2015, Published by the PCCP Owner Societies).

In conclusion, the spectra discussed here suggest new possibilities by combining SEHRS / SERS measurements of pH using NIR excitation. Both SERS and SEHRS data were excited at a wavelength of 1064 nm, and they illustrate that this combination is more powerful for the determination of micro-environmental pH (e.g., in biomaterials) and for the characterization of the sensor than one-photon spectra excited in the visible range at 532 nm alone. As shown, a combined SEHRS / SERS pH-sensor which uses *p*MBA is most sensitive in the acidic and neutral pH ranges. This makes it especially useful for the examination of biological objects, which generally profits from near-infrared excitation.

Analyzing SERS Data with Multivariate Tools

In order to achieve a separation of SERS or SEHRS reporter molecules in mixtures as well as an identification of spectra from biomolecules, multivariate methods that consider many vibrational bands are clearly of advantage. Specifically in microscopy and microspectroscopy, great potential of such a hyperspectral analysis of SERS spectra surely lies in the possibility of fingerprint-based imaging. Automated identification and classification will be crucial for localizing SERS and SEHRS nanoprobes in complex microscopic environments and for generating image contrast based on the full spectral information that they can provide.

Principal components analysis (PCA) is a multivariate technique that has found application in fields such as pattern recognition (*46*) and image compression (*47*), and has widely been applied for finding patterns in data of high dimension, in particular in hyperspectral mapping and characterization approaches (*48*, *49*).

In PCA, a data matrix is projected into a variance-weighted coordinate system. This orthogonal linear transformation (or eigenvalue problem) transforms a number of possibly correlated variables of the spectral data set into a smaller number of uncorrelated variables such that the greatest variance by any projection of the data comes to lie on the first coordinate (*50*). Thereby, in a set of SERS spectra, the different types of variance that occur, e.g., due to fluctuation of spectra, changes in signal to noise, and others, can be separated.

In the case of SERS probes, we have demonstrated the full exploitation of the multiplexing capabilities inherent to SERS nanoprobes or labels by the application of fast, multivariate methods. In a study using the SERS spectra of five reporter molecules, we have based identification of the reporter spectra on PCA and hierarchical clustering. Furthermore, data from such SERS reporter molecules were identified inside cells, in the presence of spectral contributions from the biomolecules, and used for imaging different classes of SERS spectra (*51*). Figure 7 shows such a map that uses contrast based on classification of spectral fingerprints. . In the cluster maps, each spectrum, that is, each pixel is assigned a colour, corresponding to its classification. The analysis in the data set of Figure 7 suggest that the spectra can be grouped into three major classes, one of spectra with no signals, and two other classes of SERS data.

Also in Figure 7, two maps, using the variance between the spectra of the data set represented by the first principal component and of the third principal component, respectively, are shown. To generate these maps, the first derivatives of all spectra were subjected to PCA, and the two images were recinistructed using the scores of each PC. Comparison with the cluster image reveals that

PC1, which points into the direction of greatest variance in the data set, can be used to image all spectra with SERS signal, as compared to those with no SERS signal (naturally, not all sampling volumes in the Raman map contain SERS nanoprobes, and therefore several pixels don't give a SERS spectrum). Accordingly, the loading spectrum of PC1 contains features at all frequencies where characteristic bands of the reporter molecules pABT and 2-NAT, as well as characteristic bands of cellular molecules are located. Using PC3, however, contrast can be generated between spectra containing contributions from both reporter molecules as opposed to the other spectra in the data set. Note that the locations of high scores for PC3 correspond well with the distribution of class C spectra in the cluster map.

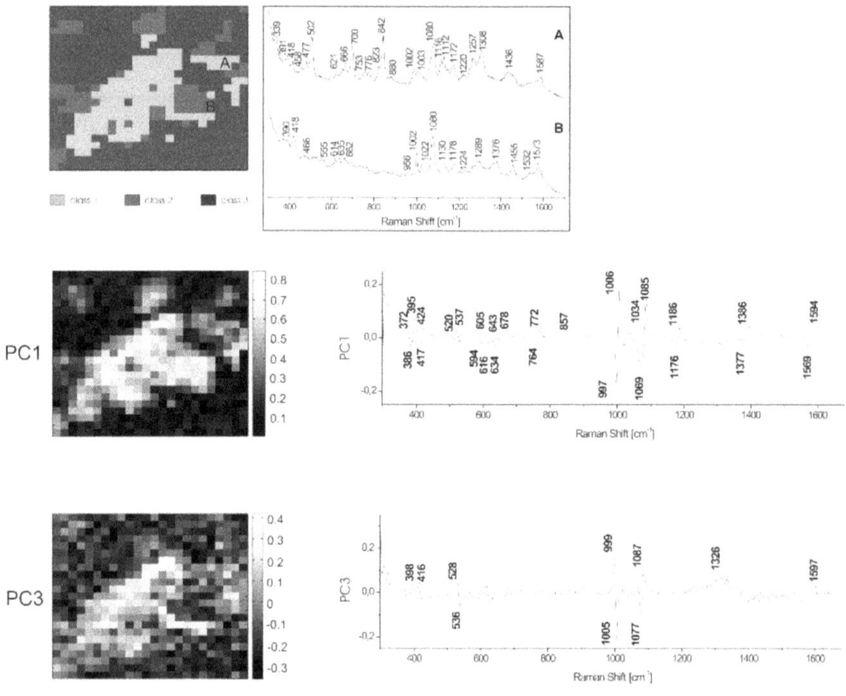

Figure 7. First line: Multivariate cluster map based on three classes and spectral information in the region 300-1700 cm^{-1}, of vector-normalized first derivative SERS spectra obtained from a fibroblast cell. Example spectra for two of the classes. HCA used Euclidean distance and Ward's algorithm. Bottom: PCA score maps of a 3T3 cell in SERS experiment with pABT and 2-NAT hybrid probes (left) and corresponding loadings spectra (right) for two PCs. The mapping data set consisted of 552 spectra. Reproduced with permission from (51). Copyright (2010, ACS).

Variation in a set of SERS data of cells or other samples can be very different compared to variation in data sets of typical normal Raman or infrared data of similar samples. While in the latter, most of the variance in the spectra is based on biochemistry, and usually covered using a limited number of principal components, e.g., five or ten, variation in SERS data can have many different origins, molecular ones, as well as variation that is inherent to the particular experiment. For example, since different nanoparticles or nanoaggregates are residing in the focal volume at the time each spectrum is acquired, the variance within the spectra from one sample can be very high. In work that was done on extracts of plant (pollen) tissue, heterogeneity in SERS data sets was studied. In liquid extracts, for example, mean spectra of several data sets, each containing 500 spectra per species were obtained from the 14 different types of samples, in order to obtain spectra that represent as many molecules contained in a preparation as possible, in spite of fluctuations that are inherent in SERS data (*52*, *53*).

Other samples and preparation procedures may require other microscopic sampling procedures in order to be able to describe the molecular heterogeneity of a material that is represented by a data set: Due to the high sensitivity of the enhancement factor signal to noise ratio in the example shown in Figure 7 varied not only between different SERS reporters, but in particular within each spectral class. In the case of the plant extracts, intra-species spectral variation due to the nature of the SERS experiment was in some cases greater than variation between species. This finding is in accordance with other recent works on the application of multivariate analyses of SERS data in bioanalytics, e.g., in diagnostics (*54*, *55*), microorganisms classification (*56*, *57*), and plant research (*58*). However, PCA can provide an approach robust enough to neglect these preparation-based differences.

To make use of this high amount of hyperspectral data, automated pattern recognition tools, in particular artificial neural networks (ANN), can also be applied. ANN are powerful for classification of large sets of spectra, and have attracted significant interest in analytical and biomedical spectroscopy. (*59–63*). The potential of ANN for fingerprint analyses of complex mixtures in biomaterials based on vibrational spectra has impressively been demonstrated by applications to bacteria (*64*), the identification of Malaria parasite cell cycles (*65*), as well as tissue imaging (*66*).

In pollen tissue, analysis of the SERS data with an artificial neural network showed that intrinsic biochemical information of the pollen cells can be utilized for the discrimination of different plant species, in spite of the great intra-species spectral variance that is caused by variations in the SERS signals and by preparation specifics (*53*). This example illustrates that ANN cannot only be exploited to identify single components in complex mixtures as shown recently (*67*), but also for the classification of complex biological extracts. In the work presented in ref. (*53*), ANN were also used to pre-select spectra that contain information relevant for classification, enabling more efficient analysis by PCA (*53*). This provides a means to use SERS data for bioanalytical mapping of other complex samples.

As suggested in recent studies on using support vector machines for SERS-based cancer (*68*) and virus (*69*) detection, the application of supervised learning based on pattern recognition to SERS data could in general be very promising.

Towards SEHRS Hyperspectral Mapping

To investigate the potential of SEHRS in multivariate mapping of large areas, in an initial study (*70*), a well-known, well-defined type of plasmonic surface, immobilized silver nanoparticles that yield homogeneous SERS enhancement on the microscopic scale (*52*), were used for first PCA-based imaging experiments. To avoid high variance due to strongly varying signal-to-noise ratio, SEHRS spectra were excited in resonance with two well-characterized reporter molecules, crystal violet (CV) and malachite green (MG). Samples were prepared with distinguishable areas for SEHRS hyperspectral mapping by immersing slides with immobilized silver nanoparticles first in MG and subsequently in CV (schematics shown in Figure 8). The region marked with the rectangle in the center is a border region between pure MG (green) and MG+CV mix (grey), and has a microscopic heterogeneity that can be investigated by SEHRS or SERS mapping. Within this border region, microscopic maps were collected.

To utilize as many as possible of the several subtle differences between both the SEHRS and the SERS spectra of CV and MG (*70*), multivariate methods were applied to map the distribution of the dyes on surfaces prepared according to the schematics shown in Figure 8.

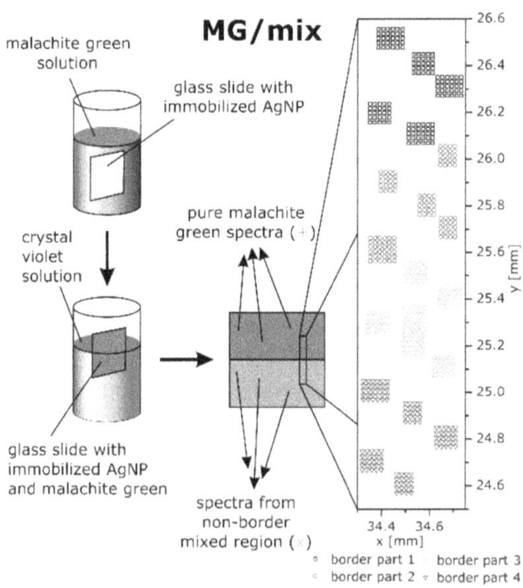

Figure 8. Schematic of the preparation of macroscopic samples that contain a border region of microscopic heterogeneity in the distribution of malachite green (MG) and crystal violet (CV) on immobilized silver nanoparticles. Reproduced with permission from (70) (Copyright 2016, Published by the PCCP Owner Societies).

For mapping, SEHRS spectra were analyzed with PCA. The scores plot in Figure 9 indicates the formation of groups of spectra that are similar with respect to a specific principal component. The loadings provide information about those bands in the spectra that are responsible for their separation with respect to the corresponding principal component. The scores of the PCA of the SEHRS spectra of an MG/mix sample yield clear separation of the MG spectra (green '+') from the mixed spectra (grey 'x') mostly due to the variation represented by the first principal component (PC) with a small contribution of the second PC. The loadings show a large variance in the spectra around 914 cm^{-1}, where the SEHRS spectra of CV and MG display bands at different frequencies (*70*). The first PC also indicates differences around 1587 cm^{-1} and 1175 cm^{-1}, where all of the SEHRS spectra show intense phenyl bands (*70*).

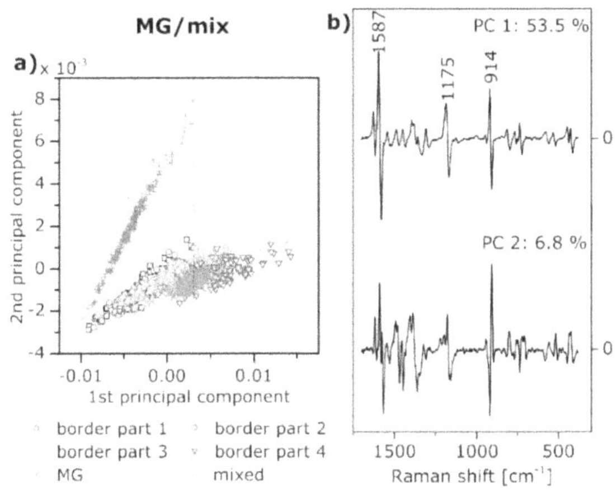

Figure 9. (a) Scores and (b) loadings of the first two principal components (PC) resulting from the principal component analysis (PCA) of the vector-normalized first derivatives of SEHRS spectra over the range 380-1700 cm^{-1} of the MG/mix samples. Symbols and colors correspond to the subregions in Figure 8. Spectra from the two non-border regions on each sample (MG and mixed, respectively) were also included in the PCA and served as references. The dashed lines in a demarcate the ranges of score values of the first PC, which were used for the differentiation between different types of spectra in Figure 10. Variance in percent that is explained by each PC is indicated in the loadings plot in b. Reproduced with permission from (70). (Copyright 2016, Published by the PCCP Owner Societies).

PCA of the SERS spectra, which were obtained from nearly the same positions as the SEHRS spectra, enables separation of the different regions of the samples in a similar way. In the scores plots of the SEHRS (Figure 9 a) and also of the the SERS data the direction of changes in the spectra from the MG/mix border regions along the first PC (Figure 9a) was used to reconstruct hyperspectral maps using the SEHRS data (Figure 10a). Similar analysis was also performed with the SERS spectra (Figure 10b). The color scale of the maps in Figure 5 was set according to the score values of the spectra, arbitrarily separating the axis of the first PC into three ranges for every scores plot. Threshold scores values for the separation are presented by the maximum and minimum scores values of the spectra of regions with only MG and mixed regions very far from the border region: The left range covers those spectra that can be unequivocally assigned to MG, the right range those from the mixed (non-border) region. In the middle, SEHRS spectra from the border, mixed, and single-analyte regions overlap.

In the upper part of the border region of the MG/mix sample, which points towards the pure MG region (Figure 10a), several SEHRS spectra are comparable to pure MG spectra (green pixels), while the rest of the SEHRS spectra and most of the SERS spectra in this part have score values within the middle range. In the lower part of the map, more spectra are assigned to the mixed region (Figure 10, grey pixels). In summary, the hyperspectral maps in Figure 10 show a gradient from the MG region to the mixed region. The differences within the border region are very small.

The separation of the score values based on the values for non-border spectra was arbitrary in this case, and needs to be defined by the respective experiment. Also, and the principal component(s) chosen for imaging are specific for each particular imaging problem (see also examples in Figure 7). For the future application of this hyperspectral imaging approach to other sample systems, the parameters for the construction of the color scale have to be determined based on appropriate reference spectra. The fact that classification of pixels in the SEHRS maps can differ from the SERS maps (Figure 10a and 10b) holds great promise to exploit the extensive chemical information of complementary SEHRS and SERS imaging.

In conclusion, we have shown simultaneous microscopic SEHRS and SERS imaging and multivariate discrimination of dyes and other molecules. For future applications, combining nonlinear and linear NIR excitation in SERS, together with the tunable optical properties of plasmonic nanoparticles, will open up new possibilities for microscopic bio-sensing. The results provide evidence that SEHRS offers additional possibilities for multivariate discrimination and mapping based on complementary vibrational information.

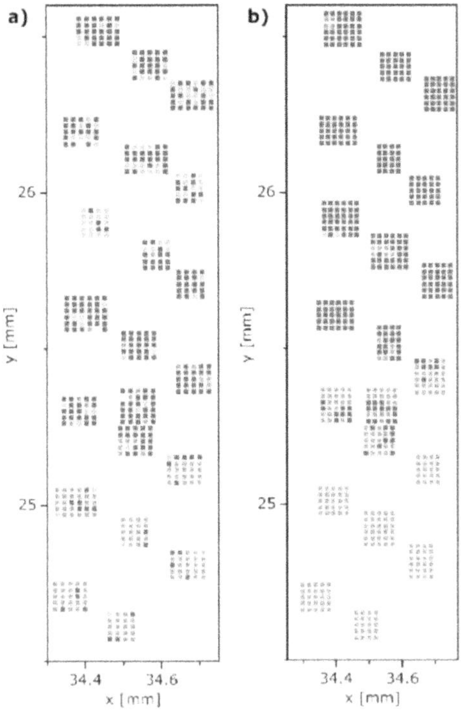

Figure 10. False-color (a) SEHRS and (b) SERS map of the border region of the MG/mix sample shown schematically in Figure 8. The maps are based on the scores of the respective first principle component of the PCA of the SEHRS spectra (Figure 9a) and of the SERS spectra (PCA not schown here). The x and y axes of the maps display the absolute coordinates of the x,y-stage that was used for the SEHRS and SERS mapping experiments in order to demonstrate the overlap of the microscopic regions that were probed by SEHRS and SERS. Color code: Assignment according to 1st PC score. Green: MG, blue: between mix and MG, grey: mix. x and y scales show absolute positions in millimeter of the x,y-stage for comparison of the two maps. Reproduced with permission from (70). (Copyright 2016, Published by the PCCP Owner Societies).

Acknowledgments

Funding of this research by ERC grant 259432 MULTIBIOPHOT is gratefully acknowledged. Zsuzsanna Heiner acknowledges funding by DFG GSC 1013 (SALSA).

References

1. Ziegler, L. D. *J. Raman Spectrosc.* **1990**, *21*, 769–779.
2. Denisov, V. N.; Mavrin, B. N.; Podobedov, V. B. *Phys. Rep.* **1987**, *151*, 1–92.
3. Cyvin, S. J.; Rauch, J. E.; Decius, J. C. *J. Chem. Phys.* **1965**, *43*, 4083–4095.
4. Kneipp, J.; Kneipp, H.; Kneipp, K. *Proc. Nat. Acad. Sci. U.S.A.* **2006**, *103*, 17149–17153.
5. Michaels, A. M.; Nirmal, M.; Brus, L. E. *J. Am. Chem. Soc.* **1999**, *121*, 9932–9939.
6. Kneipp, K.; Wang, Y.; Kneipp, H.; Itzkan, I.; Dasari, R. R.; Feld, M. S. *Phys. Rev. Lett.* **1996**, *76*, 2444–2447.
7. Nie, S.; Emory, S. R. *Science* **1997**, *275*, 1102–1106.
8. Zipfel, W. R.; Williams, R. M.; Webb, W. W. *Nat. Biotechnol.* **2003**, *21*, 1368–1376.
9. Larson, D. R.; Zipfel, W. R.; Williams, R. M.; Clark, S. W.; Bruchez, M. P.; Wise, F. W.; Webb, W. W. *Science* **2003**, *300*, 1434–1436.
10. Hulteen, J. C.; Young, M. A.; Van Duyne, R. P. *Langmuir* **2006**, *22*, 10354–10364.
11. Nie, S. M.; Lipscomb, L. A.; Yu, N. T. *Appl. Spectrosc. Rev.* **1991**, *26*, 203–276.
12. Kneipp, H.; Kneipp, K.; Seifert, F. *Chem. Phys. Lett.* **1993**, *212*, 374–378.
13. Guhlke, M.; Heiner, Z.; Kneipp, J. *Phys. Chem. Chem. Phys.* **2015**, *17*, 26093–26100.
14. Rogalski, A. *Opto-Electron. Rev.* **2012**, *20*, 279–308.
15. Rogalski, A. *Prog. Quantum Electron.* **2012**, *36*, 342–473.
16. Angel, S. M.; Katz, L. F.; Archibald, D. D.; Lin, L. T.; Honigs, D. E. *Appl. Spectrosc.* **1988**, *42*, 1327–1331.
17. Chase, D. B.; Parkinson, B. A. *Appl. Spectrosc.* **1988**, *42*, 1186–1187.
18. Crookell, A.; Fleischmann, M.; Hanniet, M.; Hendra, P. J. *Chem. Phys. Lett.* **1988**, *149*, 123–127.
19. Prucek, R.; Panacek, A.; Fargasova, A.; Ranc, V.; Masek, V.; Kvitek, L.; Zboril, R. *CrystEngComm* **2011**, *13*, 2242–2248.
20. Meyer, M. W.; Smith, E. A. *Analyst* **2011**, *136*, 3542–3549.
21. Drescher, D.; Guttmann, P.; Buchner, T.; Werner, S.; Laube, G.; Hornemann, A.; Tarek, B.; Schneider, G.; Kneipp, J. *Nanoscale* **2013**, *5*, 9193–9198.
22. Drescher, D.; Zeise, I.; Traub, H.; Guttmann, P.; Seifert, S.; Buchner, T.; Jakubowski, N.; Schneider, G.; Kneipp, J. *Adv. Funct. Mater.* **2014**, *24*, 3765–3775.
23. Buchner, T.; Drescher, D.; Merk, V.; Traub, H.; Guttmann, P.; Werner, S.; Jakubowski, N.; Schneider, G.; Kneipp, J. *Analyst* **2016**, *141*, 5096–5106.
24. Kubackova, J.; Izquierdo-Lorenzo, I.; Jancura, D.; Miskovsky, P.; Sanchez-Cortes, S. *Phys. Chem. Chem. Phys.* **2014**, *16*, 11461–11470.
25. Ansar, S. M.; Perera, G. S.; Gomez, P.; Salomon, G.; Vasquez, E. S.; Chu, I. W.; Zou, S. L.; Pittman, C. U.; Walters, K. B.; Zhang, D. M. *J. Phys. Chem. C* **2013**, *117*, 27146–27154.

26. Talley, C. E.; Jusinski, L.; Hollars, C. W.; Lane, S. M.; Huser, T. *Anal. Chem.* **2004**, *76*, 7064–7068.
27. Michota, A.; Bukowska, J. *J. Raman Spectrosc.* **2003**, *34*, 21–25.
28. Kneipp, J.; Kneipp, H.; Wittig, B.; Kneipp, K. *Nano Lett.* **2007**, *7*, 2819–2823.
29. Kneipp, J.; Kneipp, H.; Wittig, B.; Kneipp, K. *J. Phys. Chem. C* **2010**, *114*, 7421–7426.
30. Bishnoi, S. W.; Rozell, C. J.; Levin, C. S.; Gheith, M. K.; Johnson, B. R.; Johnson, D. H.; Halas, N. J. *Nano Lett.* **2006**, *6*, 1687–1692.
31. Schwartzberg, A. M.; Oshiro, T. Y.; Zhang, J. Z.; Huser, T.; Talley, C. E. *Anal. Chem.* **2006**, *78*, 4732–4736.
32. Murphy, R. F.; Powers, S.; Cantor, C. R. *J. Cell Biol.* **1984**, *98*, 1757–1762.
33. Montcourrier, P.; Mangeat, P. H.; Valembois, C.; Salazar, G.; Sahuquet, A.; Duperray, C.; Rochefort, H. *J. Cell Sci.* **1994**, *107*, 2381–2391.
34. Gühlke, M.; Heiner, Z.; Kneipp, J. *Phys. Chem. Chem. Phys.* **2015**, *17*, 26093–26100.
35. Lee, S. B.; Kim, K.; Kim, M. S. *J. Raman Spectrosc.* **1991**, *22*, 811–817.
36. Alvarez-Puebla, R. A.; Dos Santos, D. S.; Aroca, R. F. *Analyst* **2004**, *129*, 1251–1256.
37. Lee, A. S. L.; Li, Y. S. *Spectrochim. Acta, Part A* **1996**, *52*, 173–184.
38. Wang, Y.; Ji, W.; Sui, H. M.; Kitahama, Y.; Ruan, W. D.; Ozaki, Y.; Zhao, B. *J. Phys. Chem. C* **2014**, *118*, 10191–10197.
39. Ma, C. X.; Harris, J. M. *Langmuir* **2011**, *27*, 3527–3533.
40. Liu, Y.; Yuan, H.; Fales, A. M.; Vo-Dinh, T. *J. Raman Spectrosc.* **2013**, *44*, 980–986.
41. Park, H. K.; Lee, S. B.; Kim, K.; Kim, M. S. *J. Phys. Chem.* **1990**, *94*, 7576–7580.
42. Yu, Y. Y.; Handa, S.; Yajima, T.; Futamata, M. *Chem. Phys. Lett.* **2013**, *560*, 49–54.
43. Handa, S.; Yu, Y. Y.; Futamata, M. *Vib. Spectrosc.* **2014**, *72*, 128–133.
44. Alvarez-Puebla, R. A.; Arceo, E.; Goulet, P. J. G.; Garrido, J. J.; Aroca, R. F. *J. Phys. Chem. B* **2005**, *109*, 3787–3792.
45. Kazanci, M.; Schulte, J. P.; Douglas, C.; Fratzl, P.; Pink, D.; Smith-Palmer, T. *Appl. Spectrosc.* **2009**, *63*, 214–223.
46. Bereton, R. G. Pattern Recognition. In *Chemometrics: Data Analysis for the Laboratory and Chemical Plant*; John Wiley & Sons Ltd: Chichester, U.K., 2006; pp 183–269.
47. Clausen, C.; Wechsler, H. *Pattern Recognit.* **2000**, *33*, 1555–1560.
48. Koljenovic, S.; Schut, T. C. B.; van Meerbeeck, J. P.; Maat, A.; Burgers, S. A.; Zondervan, P. E.; Kros, J. M.; Puppels, G. J. *J. Biomed. Opt.* **2004**, *9*, 1187–1197.
49. Qian, X. M.; Peng, X. H.; Ansari, D. O.; Yin-Goen, Q.; Chen, G. Z.; Shin, D. M.; Yang, L.; Young, A. N.; Wang, M. D.; Nie, S. M. *Nat. Biotechnol.* **2008**, *26*, 83–90.
50. Pearson, K. *London Edinburgh Philos. Mag. J. Sci.* **1901**, *6*, 559–572.
51. Matschulat, A.; Drescher, D.; Kneipp, J. *Acs Nano* **2010**, *4*, 3259–3269.

52. Joseph, V.; Gensler, M.; Seifert, S.; Gernert, U.; Rabe, J. P.; Kneipp, J. *J. Phys. Chem. C* **2012**, *116*, 6859–6865.
53. Seifert, S.; Merk, V.; Kneipp, J. *J. Biophotonics* **2016**, *9*, 181–189.
54. Lin, D.; Feng, S.; Pan, J.; Chen, Y.; Lin, J.; Chen, G.; Xie, S.; Zeng, H.; Chen, R. *Opt. Express* **2011**, *19*, 13565–13577.
55. Feng, S.; Lin, D.; Lin, J.; Li, B.; Huang, Z.; Chen, G.; Zhang, W.; Wang, L.; Pan, J.; Chen, R.; Zeng, H. *Analyst* **2013**, *138*, 3967–3974.
56. Rivera-Betancourt, O. E.; Karls, R.; Grosse-Siestrup, B.; Helms, S.; Quinn, F.; Dluhy, R. A. *Analyst* **2013**, *138*, 6774–6785.
57. Zhou, H.; Yang, D.; Ivleva, N. P.; Mircescu, N. E.; Niessner, R.; Haisch, C. *Anal. Chem.* **2014**, *86*, 1525–1533.
58. Qiu, L.; Liu, P.; Zhao, L.; Wen, M. Q.; Yang, H. Y.; Fan, S. G.; Zhou, L. Z. *Vib. Spectrosc.* **2014**, *72*, 134–141.
59. Liu, Y.; Upadhyaya, B. R.; Naghedolfeizi, M. *Appl. Spectrosc.* **1993**, *47*, 12–23.
60. Schulze, H. G.; Blades, M. W.; Bree, A. V.; Gorzalka, B. B.; Greek, L. S.; Turner, R. F. B. *Appl. Spectrosc.* **1994**, *48*, 50–57.
61. Santos, V. O.; Oliveira, F. C. C.; Lima, D. G.; Petry, A. C.; Garcia, E.; Suarez, P. A. Z.; Rubim, J. C. *Anal. Chim. Acta* **2005**, *547*, 188–196.
62. Reisner, L. A.; Cao, A.; Pandya, A. K. *Chemom. Intell. Lab. Syst.* **2011**, *105*, 83–90.
63. Schumacher, W.; Kuehnert, M.; Roesch, P.; Popp, J. *J. Raman Spectrosc.* **2010**, *42*, 383–392.
64. Udelhoven, T.; Naumann, D.; Schmitt, J. *Appl. Spectrosc.* **2000**, *54*, 1471–1479.
65. Webster, G. T.; De Villiers, K. A.; Egan, T. J.; Deed, S.; Tilley, L.; Tobin, M. J.; Bambery, K. R.; McNaughton, D.; Wood, B. R. *Anal. Chem.* **2009**, *81*, 2516–2524.
66. Goodacre, R. *Vib. Spectrosc.* **2003**, *32*, 33–45.
67. Kasera, S.; Herrmann, L. O.; Barrio, J. d.; Baumberg, J. J.; Scherman, O. A. *Sci. Rep.* **2014**, *4*.
68. Li, S.-X.; Zeng, Q.-Y.; Li, L.-F.; Zhang, Y.-J.; Wan, M.-M.; Liu, Z.-M.; Xiong, H.-L.; Guo, Z.-Y.; Liu, S.-H. *J. Biomed. Opt.* **2013**, *18*.
69. Negri, P.; Choi, J. Y.; Jones, C.; Tompkins, S. M.; Tripp, R. A.; Dluhy, R. A. *Anal. Chem.* **2014**, *86*, 6911–6917.
70. Guhlke, M.; Heiner, Z.; Kneipp, J. *Phys. Chem. Chem. Phys.* **2016**, *18*, 14228–14233.

Editors' Biographies

Yukihiro Ozaki

Yukihiro Ozaki obtained his Ph.D. (1978) in chemistry from Osaka University. He is currently a Professor in the Department of Chemistry at Kwansei Gakuin University in Sanda, Japan. He has been active in the research of a variety of molecular spectroscopy, covering IR, Raman, NIR, and far-ultraviolet (FUV) spectroscopy. His contributions to SERS/TERS spectroscopy involve; (1) mechanism of SERS, (2) development of SERS/TERS spectrometers and experimental techniques, (3) application of TERS to nanomaterials, (4) 3-D SERS imaging, and (5) semiconductor-enhanced Raman scattering. Prof. Ozaki has received many awards including the 1998 Tomas Hirschfeld Award, the 2002 Spectroscopical Society of Japan Award, the 2005 Science and Technology Award of Japanese Government (Ministry of Education, Culture, Sports, Science and Technology), the Gerald Birth Award, and the Bomen-Michelson Award. He is also a fellow of the Royal Society of Chemistry, the Chemical Society of Japan, and the Society for Applied Spectroscopy.

George C. Schatz

George C. Schatz is Charles E. and Emma H. Morrison Professor of Chemistry and of Chemical and Biological Engineering at Northwestern University. He received his undergraduate degree in chemistry at Clarkson University and a Ph.D at Caltech. He was a postdoc at MIT and has been at Northwestern since 1976. Schatz has published three books and more than 800 papers. Schatz is a member of the National Academy of Sciences, the American Academy of Arts and Sciences, the International Academy of Quantum Molecular Sciences, and he has been Editor-in-Chief of *The Journal of Physical Chemistry* since 2005. Recent awards include the Debye and Langmuir Awards of the ACS, the S F Boys-A Rahman Award of the Royal Society of Chemistry, and the Hirschfelder Award of the University of Wisconsin. He is a Fellow of the APS, RSC, ACS, and the AAAS. Schatz is a theoretician who studies the optical, structural, and thermal properties of nanomaterials, including plasmonic nanoparticles, DNA and peptide nanostructures, and carbon-based materials, with applications in chemical and biological sensing, electronic and biological materials, high performance fibers, and solar energy. His past work has also been concerned with understanding the dynamics of chemical reactions in the gas phase and in gas-surface collisions.

Duncan Graham

Duncan Graham is Research Professor of Chemistry and Head of Department for Pure and Applied Chemistry at the University of Strathclyde in Glasgow, having being appointed there as a lecturer in 2002 and then as a Chair in 2004. He is currently Chair of the Editorial Board of *Analyst* and President elect of the Analytical Division of the Royal Society of Chemistry. He serves on the advisory boards of *Chemical Society Reviews, Chemical Science, Journal of Raman Spectroscopy*, the journal of *Biomedical Spectroscopy and Imaging* and the new Cell Press journal, *Chem*. He has been awarded numerous awards for his research including the RSC's SAC Silver Medal (2004), the RSC's Corday Morgan Prize (2009), a Royal Society Wolfson Research Merit Award (2010), the Craver Award from the Coblentz Society (2012), and the Fellows Award from the Society for Applied Spectroscopy (2012). He was elected to the fellowship of the Royal Society of Edinburgh in 2008. He has published more than 200 papers and graduated more than 50 Ph.D. students. He is a Cofounder and Director of Renishaw Diagnostics Ltd. (2007), which now has 42 FTE and has a CE marked SERS- based diagnostic now available for use in European hospitals. He completed a Ph.D. in organic chemistry at the University of Edinburgh (1996), and his interests are in developing new diagnostic assays based on nanoparticles and spectroscopy with target molecules including DNA, RNA, proteins, and small molecule biomarkers.

Tamitake Itoh

Tamitake Itoh is currently a Senior Researcher of the National Institute of Advanced Industrial Science and Technology (AIST) in Japan. He completed his Ph.D. (2002) at Osaka University under the supervision of Prof. Hiroshi Masuhara and Prof. Tsuyoshi Asahi. Through postdoctoral research positions from 2002 to 2005, he worked at Kwansei Gakuin University. In April of 2005 he became a Researcher in AIST, and in April of 2010 he was promoted to Senior Researcher. In 2012, as an additional position he was appointed as visiting Associate Professor at Nagoya University. His research theme is surface enhanced spectroscopy at ensemble- and single plasmonic nanoparticle levels.

Indexes

Author Index

Subject Index

M

P

S

www.ingramcontent.com/pod-product-compliance
Ingram Content Group UK Ltd.
Pitfield, Milton Keynes, MK11 3LW, UK
UKHW020813290926
15470UKWH00003B/1138

9 780841 232013